Proton Exchange Membrane Fuel Cells

Alhussein Albarbar • Mohmad Alrweq

Proton Exchange Membrane Fuel Cells

Design, Modelling and Performance Assessment Techniques

Springer

Alhussein Albarbar
School of Engineering
The Manchester Metropolitan University
Manchester, United Kingdom

Mohmad Alrweq
School of Engineering
The Manchester Metropolitan University
Manchester, United Kingdom

ISBN 978-3-319-70726-6 ISBN 978-3-319-70727-3 (eBook)
https://doi.org/10.1007/978-3-319-70727-3

Library of Congress Control Number: 2017957605

Printed on acid-free paper

This Springer imprint is published by Springer Nature
The registered company is Springer International Publishing AG
The registered company address is: Gewerbestrasse 11, 6330 Cham, Switzerland

Preface

This book examines the characteristics of proton exchange membrane (PEM) fuel cells with a focus on deriving realistic finite element models. This book also explains, in detail, how to set up measuring systems, data analysis and PEM fuel cells' static and dynamic characteristics. Covered in detail are design and operation principles such as polarisation phenomenon, thermodynamic analysis and overall voltage; failure modes and mechanisms such as permanent faults, membrane degradation and water management; and modelling and numerical simulation including semi-empirical, one-dimensional, two-dimensional and three-dimensional models. It is appropriate for graduate students, researchers and engineers who work with the design and reliability of hydrogen fuel cells, in particular proton exchange membrane fuel cells.

This book comes in seven chapters and is organised as follows:

Chapter 1 introduces the role of fuel cells in today's and future power (energy) systems, their types and technologies. Advantages and disadvantages of each type in addition to their applications are also covered in this chapter.

Chapter 2 presents an up-to-date survey on work carried out in the field of modelling and reliability analysis for PEM fuel cells and energy systems. This chapter covers all the literature regarding performance evaluation and efficiency calculation related to PEM fuel cells. In addition, it concentrates on reviewing the causes of the degradation in FCs and the diagnosis and monitoring techniques currently in use.

Chapter 3 explains the physical construction of PEM fuel cells as well as studies their characteristics and polarisation curves. In addition, the dynamic behaviour of the PEM fuel cells (thermodynamic) and efficiency of the PEM fuel cells are also covered in this chapter.

Chapter 4 illustrates the failure modes that occur within PEM fuel cells and their effect on the performance, reliability and efficiency of fuel cells. Additionally, this chapter also covers fault mitigation strategies that are used to reduce the negative impact of common failure modes. Finally, some diagnosis and monitoring techniques are also presented.

Chapter 5 proposes realistic finite element model based upon fundamental theory of PEM fuel cells. Simulink models to evaluate the effects of parameter variations such as temperature, pressure and water content on the performance and reliability of the PEM fuel cells are also presented.

Chapter 6 explains the experimental set-up constructed to study and investigate the effect of loading conditions on the PEM fuel cell characteristics. This includes electric and dynamic characteristics such as acoustic emission under flooding conditions. Thorough discussions and comments on those results are also provided.

Chapter 7 shows, in a step-by-step style, the derivation of models used in this book – finite element model and codes written in MATLAB and Simulink. The construction of finite element models for single PEM fuel cell is explained in detail. MATLAB and Simulink models that include typical modelling of PEM fuel cell characteristics are also demonstrated.

Manchester, UK Alhussein Albarbar
Manchester, UK Mohmad Alrweq

Contents

Abbreviations

A	Ampere
AE	Acoustic emission
AFC	Alkaline fuel cell
Bp	Bipolar plate
CHP	Combined heat and power
CTE	Coefficients of thermal expansion
CL	Catalyst layer
CO_2	Carbon dioxide
D_f	Diffusion flux of reactant species
E_o	Open-circuit voltage
F	Faraday constant
FC	Fuel cell
GDL	Gas diffusion layer
GL	Gas channel
H_2	Hydrogen
H_2O	Water
HFCs	Hydrogen fuel cells
HO•	Peroxide
HOO•	Hydroperoxide
HP	Hydropower
HRES	Hybrid renewable energy system
Hrs	Hours
I	Current density (A/cm^2)
i_a	Current density at anode side (A/cm^2)
I_c	Current density at cathode side (A/cm^2)
I_{Cell}	Actual cell current
i_o	Exchange current density
M	Molecular weight of reactant species
MCFC	Molten fluid carbonate fuel cell
MEA	Membrane electrode assembly

N	The number of electrons
Ni	Nickel
O.C	Open circuit
O_2	Oxygen
PAFC	Phosphoric acid fuel cell
PEM	Proton exchange membrane
PEM fuel cell	Proton exchange membrane fuel cell
PFSA	Perfluorinated sulfonic acid
P_{H2}	Pressure of hydrogen
P_{input}	Power input
P_{O2}	Pressure of oxygen
P_{out}	Power output
Pt	Platinum
PTFE	Polytetafluor-ethylene
PV	Photovoltaic
R	Universal gas constant, 8.314 (J/mol K)
R_e	Conductive FC components
RHE	Reversible hydrogen electrode
R_i	Ionic resistance of the electrolyte
Ru	Ruthenium
S	Source term
SOFC	Solid oxide fuel cell
T	Temperature (K)
T_a	Tantalum
T_i	Titanium
u	Velocity vector
UPS	Uninterrupted power supply
V	Voltage
WT	Wind turbine
Z_r	Zirconium
E	Reverse voltage
P	Pressure
k	Permeability of the porous medium

Greek Symbols

ηohm	Ohmic polarisation
η_{act}	Activation polarisation
η	Cell efficiency
$\zeta_{thermal}$	Thermal efficiency
δ_{thick}	The thickness of the membrane
σ	The conductivity
η_{conc}	Concentration polarisation
Ω	Ohm
η_{thermal}	Thermal efficiency
ΔG_f	Gibbs energy
Δg_f	Enthalpy
α	Charge transfer coefficient
ψ	Fuel utilisation efficiency

Chapter 1
Introduction and Background

1.1 Overview

Costs of fossil fuel and the growing concerns about greenhouse gas emissions have prompted scientists and researchers to look for other sources of energy that are sustainable with low costs, highly efficient energy conversion and minimal environmental impact. Therefore, interest in green renewable power generation systems, such as photovoltaic (PV), wind turbine (WT), hydropower (HP) and hydrogen fuel cells (HFCs), has increased [1–5]. Although PV, WT and HP renewable systems represent effective ways to substantially reduce fuel consumption and emissions, these systems have their own drawbacks and limitations. For example, solar and wind energy systems experience large variations in their output power since they mainly rely on weather conditions, while hydropower depends on geographic position and water availability [6]. Usually PVs and WTs are combined together with the use of an energy-storage system, such as batteries and fuel cells, as a type of energy-balancing medium. This system is known as the hybrid renewable energy system (HRES), and it is employed to overcome the drawbacks of PVs and WTs [1, 7]. Unlike other renewable energy power generations, FCs do not rely on weather conditions or geographic position.

Hydrogen fuel cells (HFCs) are part of the energy economy, which are a renewable resource when produced from water electrolysis using electricity from solar, wind and bioenergy sources (see Fig.1.1). Hydrogen is an attractive alternative since water is a by-product when it reacts with air in a fuel cell. Therefore, hydrogen and FC technologies can address two essential challenges, which are reducing CO_2 emissions and reducing overall dependence on fossil fuels.

A. Albarbar, M. Alrweq, *Proton Exchange Membrane Fuel Cells*,
https://doi.org/10.1007/978-3-319-70727-3_1

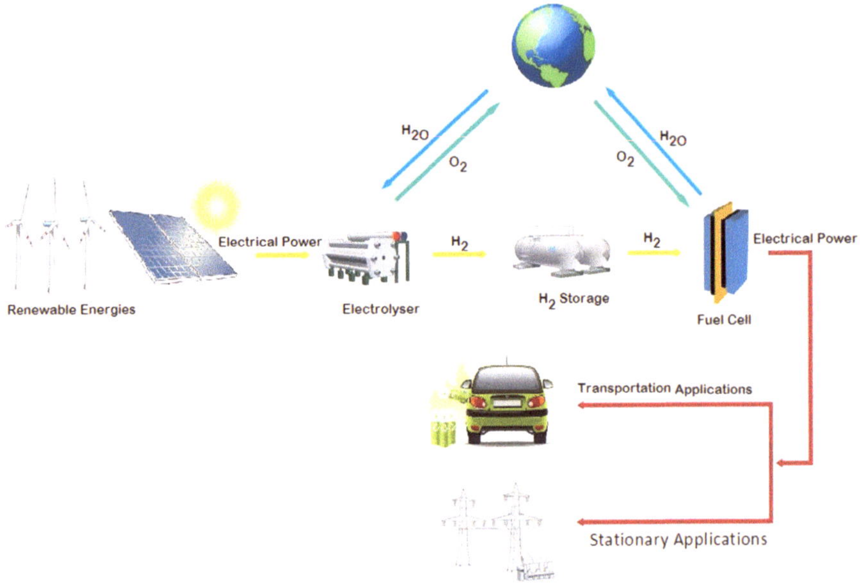

Fig. 1.1 Renewable energy, fuel cell and hydrogen economy

1.2 Fuel Cells Technologies

A fuel cell is an electrochemical device that converts chemical energy into electrical energy. Due to the electrochemical reactions occurring on both electrodes of the FC, the fuel at the anode side is oxidised to release electrons, which are then transferred to the cathode side via an external circuit, reducing the oxidant species (usually oxygen). The flow of electrons during these electrochemical processes gives rise to current in the electrical circuit, while potential differences exist over the two electrodes based upon the nature of the redox reactions. The electrical energy obtained by FCs can be utilised in many sectors such as in residential, stationary, cell phone and communication and transport applications. The greatest advantage of FC energy technologies compared to the traditional ones is their high efficiency. Unlike internal combustion and steam engines, heat exchange and mechanical work are no longer the major energy conversion methods. The electrons from the chemical reactions themselves are collected and converted directly to supply power. FC is a clean source of energy if proper selection of fuel, such as pure hydrogen, has been used, showing great potential of mitigating the environmental pollution problems of the modern industrial world. State-of-the-art FCs are appearing on the market in two forms, stand-alone or networked, to certain residential and industrial applications or as an uninterrupted power supply (UPS), as a major power system for adventure and telecommunications in distant locations and as demonstration units in universities and automobile companies.

Over the last 50 years, much research has been conducted to develop FC products due to their high efficiency and environmentally friendly operation. However, the results have only slightly improved performance. The durability and reliability have shown no major benefits. FCs have not been widely adopted in the open market due to their low durability and reliability and unacceptably high cost. Those weaknesses remain strong research issues and appear to be the most important considerations in achieving successful commercialisation of FCs (targets: 40,000 h for stationary applications and about 5000 h for automotive) [8, 9].

1.2.1 Fuel Cell Types and Applications

1.2.1.1 Fuel Cell Types

Fuel cells can be classified according to the type of fuel used, which can be pure hydrogen or gases saturated with hydrogen, and also depending on the specified reaction temperature where some FCs work at low temperatures (less than 100 °C) such as the proton-exchange membrane (also known as the polymer electrolyte membrane fuel cell) (PEM fuel cell), while others work at high temperatures (~600–1000 °C) such as solid oxide fuel cells (SOFCs).

In this research, the classification adopted is one based on the type of material used for the electrochemical material (i.e. electrolyte), and their types according to this classification are:

1. Alkaline fuel cells (AFCs)
2. Solid oxide fuel cells (SOFCs)
3. Phosphoric acid fuel cells (PAFCs)
4. Melton fluid carbonate fuel cells (MCFCs)
5. Proton-exchange membrane fuel cells (PEMFCs)

Table 1.1 briefly lists some of the properties of these cells.

1.2.1.2 Fuel Cell Applications

Based on each FC type's characteristics, FCs may have specific applications in three main categories: transportation, stationary and portable applications. For instance, FCs have been used as the auxiliary power supply in the Apollo and Gemini space programmes, respectively. FCs may become the future replacement of current internal combustion engines in automobiles. High-temperature FCs such as MCFCs and SOFCs have been applied in stationary power stations to generate electricity and heat for the community. In addition, researchers are investigating the possibility of DMFCs as the power supplier for portable electronic apparatus including cell phones and laptops. In addition, PEM fuel cells have been used in transit utility vehicles, breath alcohol testers and outdoor activity power sources [10–12].

Table 1.1 Types of fuel cells and their characteristics and usages

Proton-exchange membrane fuel cell (PEM fuel cell)	Membrane	Solid polymer
	Operating temperature	(20–180) °C
	Anode reaction	$H_2 \rightarrow 2H^+ + 2e^-$
	Cathode reaction	$½O_2 + 2H^+ + 2e^- \rightarrow H_2O + heat$
	Mobile ion	H^+
	Stack size	<1–100 Kw
	Efficiency	60% transportation 35% stationary
	Applications	Portable power Transportation Stationary power
	Advantages	Solid membrane Low temperature Quick start-up
	Disadvantages	Expensive catalyst Sensitive to fuel impurities
Alkaline fuel cell (AFC)	Membrane	Solution (KOH)
	Operating temperature	60–120 °C
	Anode reaction	$H_2 + 2(OH)^- \rightarrow H_2O + 2e^-$
	Cathode reaction	$½O_2 + H_2O + 2e^- \rightarrow 2(OH)^-$
	Mobile ion	$(OH)^-$
	Stack size	(10–100) kw
	Efficiency	60%
	Applications	Military uses Spacecraft
	Advantage	High performance due to the fast electrochemical reactions Low cost of components
	Disadvantage	Sensitivity to carbon oxide poisoning Electrolyte management
Phosphoric acid fuel cell (FAFC)	Membrane	Phosphoric acid
	Operating temperature	160–200 °C
	Anode reaction	$H_2 \rightarrow 2H^+ + 2e^-$
	Cathode reaction	$½O_2 + 2H^+ + 2e^- \rightarrow H_2O$
	Mobile ion	H^+
	Stack size	400 Kw
	Efficiency	85% combined heat and power (CHP) 37–42% (generation of power)
	Applications	Distributed generations
	Advantages	Higher temperature Tolerant of fuel impurities
	Disadvantages	Pt catalyst Long start-up time

(continued)

Table 1.1 (continued)

		Large and heavy Expensive Low output power
Molten carbonate fuel cell (MCFC)	Membrane	Molten carbonate
	Operating temperature	500–650 °C
	Anode reaction	$H_2 \rightarrow 2H^+ + 2e^-$
	Cathode reaction	$½O_2 + 2H^+ + 2e^- \rightarrow H_2O$
	Mobile ion	H^+
	Stack size	0.3–3 MW
	Efficiency	45–50%
	Applications	Distributed generations
	Advantage	Higher temperature Tolerant of impurities fuel
	Disadvantage	Corrosion and breakdown of cell components Long start-up time Low current density
Solid oxide fuel cell (SOFC)	Membrane	Ceramic compound
	Operating temperature	600–1000 °C
	Anode reaction	$H_2 + O^- \rightarrow H_2O + 2e^-$ $CO + O^- \rightarrow CO_2 + 2e^-$ $CH_4 + 4O^- \rightarrow 2H_2O + CO_2 + 8e^-$
	Cathode reaction	$½O_2 + CO_2 + 2e^- \rightarrow CO_3^{-2}$
	Mobile ion	O^-
	Stack size	1–3 MW
	Efficiency	85% combined heat and power (CHP) 60% (generation of power)
	Applications	Auxiliary power Electrical utility Distribution generation
	Advantage	High efficiency Fuel flexibility Variety of catalysts Solid membrane
	Disadvantage	High temperature Corrosion and breakdown of cell components Long start-up time

1.2.2 Full Cell Advantages

Fuel cells can be considered as a battery, except that a fuel cell is easily refuelled since the fuel is stored outside the cells. In fuel cells, fuel constantly flows through the cell, and it is easily replenished outside the cell, so it never stops providing electrical current. If there is a flow of fuel through the cell, the electricity flows out of it. A fuel cell is an energy-producing device, while a battery is an energy-storage device. For batteries, the fuel is stored inside the battery and cannot easily be replenished. Another unique feature that FCs possess is the ability to scale between power (specified via FCs' size) and capacity (specified by reservoir size). Batteries cannot scale well at large sizes. This is not applied for FCs that scale well from 1 W range to megawatt ranges. Moreover, FCs provide energy with high densities compared to batteries. In addition, FCs can be quickly recharged via refuelling, while batteries are not rechargeable or need a long time to be recharged [11].

Unlike the conventional energy conversion through the combustion process, fuel cells convert the chemical energy of the fuel directly to electrical energy, which is not subject to Carnot thermodynamic inefficiency. Fuel cells possess 45–60% efficiency compared with the 33% efficiency of a fossil fuel plant, as shown in Fig. 1.2. Fuel cells allow conversion of a wide range of gasifier hydrocarbon fuels such as ammonia, natural gas, gasified coal, etc. Unlike conventional combustion-based power generation technologies, FCs can significantly decrease any detrimental environmental impact [12, 13]. The pollution reductions can be seen from Fig. 1.3, where PEM fuel cell emits virtually gas emissions, and the emitted CO_2 is significantly reduced by more than 50%.

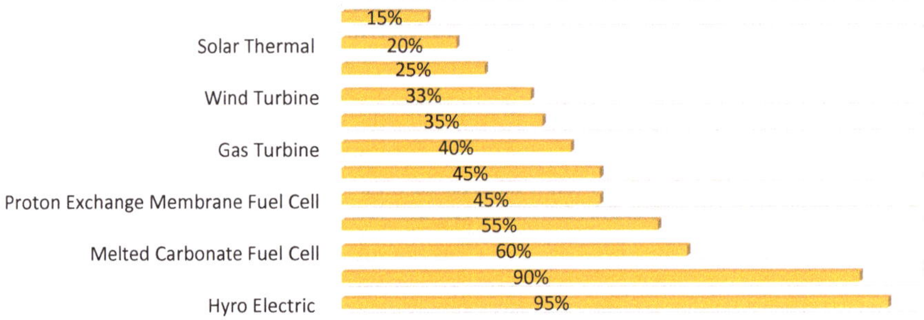

Fig. 1.2 Comparison of the efficiency of different sources of electrical power [7]

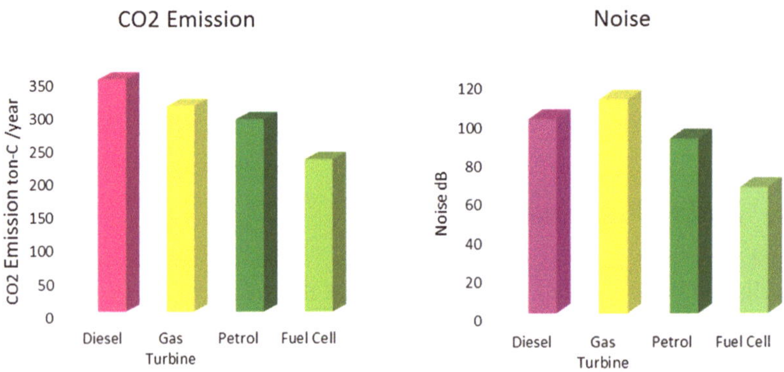

Fig. 1.3 Fuel cell emission environmental benefits [7]

1.2.3 Full Cell Disadvantages

Despite the advantages that FCs have, they also possess several disadvantages. The first one is the implementation cost of FCs, which limits FC application. Although there are great efforts to bring FCs into the energy markets, the high cost remains an obstacle.

Power density is the power produced by FC per volume or per unit mass. Despite recent improvements in power density, more improvements are needed for FCs to be applied in portable and automotive applications. Generally, engines and batteries are more effective than FCs in terms of volumetric power density. However, in terms of gravimetric power density, FCs, batteries and engines have closed performance.

Further two issues are availability and storage. FCs work perfectly with hydrogen gas, which is not widely available, and the source of pure hydrogen remains an unresolved issue for PEM fuel cell application. Although hydrogen can be obtained from many sources, for example, water electrolysis, methanol/natural gas/gasoline reformation and bacterial production, the cost of hydrogen generation, storage and building hydrogen refuelling infrastructure is high.

Reformation of hydrocarbon compounds cannot eliminate the dependence of power generation on naturally conserved fossil resources and has carbon dioxide (CO_2) as one of the products, causing environmental issues by itself. Moreover, they gave low volumetric power density, and these FCs are hard to store.

Alternative fuels such as gasoline and formic acid are difficult to use directly, and they usually need reformation. These problems reduce the FC performance. Gasoline is a good alternative fuel from the energy density point of view; however, it is not suitable for FC use.

Some additional limitations of FCs include the operational temperature compatibility issues, lack of environmental safety and durability [14]. The next chapter of the book investigates the performance of proton-exchange membrane (PEM) fuel cells and outlines methods to enhance their performance and efficiency.

References

1. Rodrigues, S., Restrepo, C., Kontos, E., Pinto, R. T., & Bauer, P. (2015). Trends of offshore wind projects. *Renewable and Sustainable Energy Reviews, 49*, 1114–1135.
2. Albarbar, A., & Batunlu, C. (2018). Introduction. In: *Thermal analysis of power electronic devices used in renewable energy systems*. Cham: Springer, ISBN 978–3–319-59827-7. https://doi.org/10.1007/978-3-319-59828-4_1.
3. Alrweq, M., & Albarbar, A. (2016). Investigation into the characteristics of proton exchange membrane fuel cell-based power system. *IET Science, Measurement and Technology*. https://doi.org/10.1049/iet-smt.2015.0046. Online ISSN 1751-8830.
4. Natsheh, E., Natsheh, A., & Albarbar, A. (2013). Intelligent controller for managing power flow within standalone hybrid power systems. *IET Science, Measurement and Technology, 7* (4), 191–200.

5. Batunlu, C., & Albarbar, A. (2016). Real-time system for monitoring the electro-thermal behaviour of power electronic devices used in boost converters. *Microelectronics Reliability, 62*(2016), 82–90.
6. International Energy Agency. (2015). *Key world energy statistics 2015*. http://www.iea.org/publications/freepublications/publication/KeyWorld_Statistics_2015.pdf.
7. Panwar, N., Kaushik, S., & Kothari, S. (2011). Role of renewable energy sources in environmental protection: A review. *Renewable and Sustainable Energy Reviews, 15*(3), 1513–1524.
8. Kumar, A., Sharma, S., Kumar, K., Kaushik, N., & Mishra, S. (2010). Renewable energy in India: Current status and future potentials. *Renewable and Sustainable Energy Reviews, 14*(7), 2434–2442.
9. Kirsten, S. (2014). Renewable energy sources act and trading of emission certificates: A national and a supranational tool direct energy turnover to renewable electricity-supply in Germany. *Energy Policy, 64*, 302–312.
10. Bose, S., Kuila, T., Nguyen, H., et al. (2011). Polymer membranes for high temperature proton exchange membrane fuel cell: Recent advances and challenges. *Progress in Polymer Science, 36*(6), 813–843.
11. Dutta, S., Shimpalee, S., & Van Zee, W. (2001). Numerical prediction of mass-exchange between cathode and anode channels in a PEM fuel cell. *International Journal of Heat and Mass Transfer, 44*(11), 2029–2042.
12. Murray, E., Perry, T. T., & Barnett, S. A. (1999). A direct-methane fuel cell with a ceria-based anode. *Nature, 400*(6745), 649–651.
13. Bianchi, F. D., et al. (2015). Fault-tolerant unfalsified control for PEM fuel cell systems. *IEEE Transactions on Energy Conversion, 30*(1), 307–315.
14. Jouin, M., Gouriveau, R., Hissel, D., Péra, M. C., & Zerhouni, N. (2014). Prognostics of PEM fuel cell in a particle filtering framework. *International Journal of Hydrogen Energy, 39*(1), 481–494.

Chapter 2
Proton Exchange Membrane Fuel Cells: Review

2.1 Overview

A considerable body of research work has been carried out to develop highly efficient and reliable single PEM fuel cell and stacks to be utilised in various applications such as portable devices as a power source, stationary and automotive applications. Significant development has been achieved recently, especially in materials and current density, which will eventually lead to increasing the power density and make the device more efficient and reliable. It was expected that proton exchange membranes (PEM) fuel cells would be commercialised for static applications by 2001 and in transport applications in 2003 [1]. However, some of technical challenges remain as obstacles in the way to this target. Namely, those are related to the infrastructure and storage of hydrogen fuel in addition to the fuel cell systems. One of the major issues is the high cost of the fuel cell itself. Other technical barriers to adopting the technology, in practical as power sources, are the performance, reliability and durability under different operating conditions. The requirements for fuel cell lifetime vary significantly, per the nature of the application, ranging from 5000 h for cars to 20,000 h for buses and 40,000 h for static applications with continuous operation. Despite the fuel cell's targeted age, in transportation sector, these are much lower than for static applications, the operating conditions such as the operations of start-up and shutdown, dynamic load including load cycling and freeze–thaw make this target very challenging for current available technologies.

Fabrication and assembly of fuel cells, as well as the operational conditions, control strategies, degradation of materials, impurities and contaminants are the major concerns that must be considered when evaluating their performance and reliability.

The following terms and definitions should be clearly outlined in order to understand the concepts of PEM fuel cells performance deterioration and lifetime:

A. Albarbar, M. Alrweq, *Proton Exchange Membrane Fuel Cells*,
https://doi.org/10.1007/978-3-319-70727-3_2

- Reliability: the capability of fuel cells to accomplish the required functions while working under normal operating conditions, within the defined run time. That covers the failure modes, which cause a major reliability issue and performance under the acceptable level.
- Durability: the capability of a fuel cell to resist performance degradation over time. Degradation in durability leads to an unrecoverable failure. This accountability is closely linked to the operational cell's age.
- Stability: the capability of a fuel cell to recover loss of energy while the cell is in operating. Stability decay is closely associated with the operating conditions of the device, such as poor water and heat management.

The reliability and performance of a fuel cell's overall deterioration rate are usually measured when it is in operation (in a continuous condition) without any interruption for a specific time. Performance of the fuel cell can be expressed as the sum of the durability and stability decay rates.

Throughout this chapter, studies made by academic and industrial researchers on the performance, efficiency and reliability for PEM fuel cells have been summarised. It includes the operating conditions of the fuel cell at steady state or at accelerated conditions. Steady state and accelerated conditions include air and/or fuel starvation load or thermal cycles.

2.2 Lifetime Tests for Proton Exchange Membrane Fuel Cells

Several studies have been conducted to understand the deterioration mechanics of the components of PEM fuel cells such as membranes, catalyst layers (CLs), bipolar plates (Bps), gas channels (GLs) and gas diffusion layers (GDLs). However, only a few studies have focused on improving the reliability and lifetime of PEM fuel cells due to several reasons such as high cost and long duration of any realistic tests. Recent studies stated that for a PEM fuel cell bus system, an uninterrupted testing is needed for approximately 4.5 years at a cost of ~$2 M [2, 3].

Increasing the sample productivity and reducing the testing time required are the targets of many fuel cell designers and companies, such as General Motors, Gore, DuPont and Ballard Power Systems. To achieve this goal, various accelerated stress tests (ASTs) have been suggested and carried out to set the reliability, durability and performance of present fuel cell components. Most of the experiments on fuel cells' lifetime under steady-state conditions were conducted for much less than 4000 h (despite demonstrating a degradation rate between 2 and 10 μVh^{-1}).

To commercialise fuel cells, the degradation rate must be less than 10 μVh^{-1}. If the cell degradation rate exceeds this value (10 μVh^{-1}), additional investigations on the components and the failure mode system are required.

2.3 A Review of Performance Degradation and Common Failure Modes

The faults that occur inside fuel cell are many, ranging from chemical faults, electrochemical, electrical and mechanical faults. This section covers most of the faults that have significant impact on the performance and reliability of PEM fuel cells.

2.3.1 Membrane Degradation

The membrane (electrolyte) is one of the essential and expensive components in PEM fuel cells. In a typical PEM fuel cell, the membrane is placed between two catalyst layers of electrodes. The membrane allows protons to pass from negative electrode (anode) to positive electrode (cathode) and separates oxidising gases at the anode and reduces gases at the cathode. To ensure the achievement of all these functions, membrane should be of excellent thermal stability, ionic conductivity and chemical and flow reactant gas permeability [4–6]. Nafion® (Dupont TM), Aciplex®, Flemion® (Asahi TM) and Gore-Select® (Gore TM) membranes are most commonly used in PEM fuel cells. Several studies have been conducted on the operation/mechanism of the membranes decay and failure in the fuel cells. Performance, reliability, efficiency and durability of the polymer exchange membrane used for PEM fuel cells are areas still under continuous assessments.

2.3.1.1 Membrane Degradation Mechanisms

Air and hydrogen flow rates, through both sides of the PEM fuel cells, are comparatively slow and result in a loss of efficiency of about 1–3% [7, 8]. However, the exothermal combustion of H_2 and O_2 can cause pinholes in the membrane and distresses which may lead to damage of membrane electrode assembly (MEA). Under severe conditions, the chemical reaction that takes place at the catalyst layers of both electrodes can produce hydroperoxide and peroxide radicals, which cause chemical damage to the membrane and catalysts [9–11]. It was also reported that the rate of production of these radicals and the resulting chemical destruction of the membrane become more accelerated at low humidity rates and at open-circuit voltage [12, 13]. Some recent publications have confirmed that the ionic group's loss starts on the negative side of the membrane and eventually advances towards the positive side [14, 15]. However, others [16, 17] mentioned it also occurs in the cathode side. On the other hand, Subban et al. [18] found no noticeable difference between the negative side and the positive side of the membrane. The potential sources of multivalent ion contaminants are many, such as impurities in the air stream and corrosion of stack components, humidifier reservoirs, etc. [19–21]. A greater number of active sites were filled by the multivalent ions during the operation of the fuel cells. As a result, water molecules and H^+ transferred ions are affected by the change of membrane ionic conductivity [22]. This has no severe impact under normal conditions, unless the concentration of contamination exceeds 50% of the sulfonic acid sets in the PEM fuel cells [23].

The other probable mode of degradation is caused by the changed in water flux inside the membrane. A small level of contaminants (<5%) is sufficient for damage to take place. The proton conductivity and attenuation of water flux are also caused by the movement of H^+ with foreign particles, which in turn result in much more rapid and extensive membrane drying, particularly near the negative electrode [23, 24].

Trace metal ions, produced because of the corrosion, such as $Cu^{2+}$and Fe^{2+}, can result in contamination, accelerate membrane thinning rate and degrade the performance of PEM fuel cell. The process can be expressed in the Eqs. 2.1, 2.2, 2.3 and 2.4 [24]:

$$H_2O_2 + Fe^{2+} \rightarrow HO^{\bullet} + HO^- + Fe^{3+} \quad (2.1)$$

$$Fe^{2+} + HO^{\bullet} \rightarrow Fe^{3+} + HO^- \quad (2.2)$$

$$H_2O_2 + HO^{\bullet} \rightarrow HO^{\bullet}{}_2 + H_2 \quad (2.3)$$

$$Fe^{2+} + HO_2\cdot \rightarrow Fe^{2+} + H^+ + O_2 \quad (2.4)$$

This mechanism could result in the occurrence of pinholes or thinning in the membrane and subsequently cause an irreversible failure of the PEM fuel cell. According to the type of the membrane, radicals produced during the reaction can displace the aromatic group in carbon, such as the branching points of the electrolyte (polymer). On the other hand, some H-containing terminal bonds, with carboxylate end groups, are inexorably created during the production process of polymer and are considered the inducing agent for the chemical degradation of membranes that have high susceptibility to radical attacks. The unzipping reaction is a commonly accepted procedure that initiates the releasing of hydrogen from the end groups, free hydrogen fluoride (FHF) and carbon dioxide (CO_2) and creates new carboxylate groups at the chain ends [25].

The radical attack might circulate along the main chain as the process repeats and ultimately lead to the decomposition of the polymer into low-molecular weight compounds. Jeon et al. [26] suggested another probable mechanism which is the separation of the main chains of polymer, where the ether linkages are considered to be the most targeted side chain sites for attack by radicals, generating vulnerable –COOH groups. Consequently, the number of –COOH groups increased, while the average molecular weight of the polymers decreased together with time. Even in the absence of susceptible end groups, the polymer chain of the PFSA membrane might selectively interact under exposure to H_2.

2.3.1.2 Mitigation Techniques for Membrane Degradation

To avoid or mitigate the mechanical failure of the proton exchange membrane (PEM), the membrane electrode assembly (MEA) and flow field structure of the cell should be carefully designed. In addition, preventing the hydration levels of the membrane to lower the allowable limit is essential, specifically at the reactant inlet area. Gore FC Technologies developed a PEM reinforced with e-PTFE, which demonstrated a lifetime, much longer than a non-reinforced PEM of relative thickness. Koh et al. [27] reported similar results for PEM with enhanced mechanical strength through use of 'Nafion®–Teflon®–phosphotungstic acid' composite membranes and reinforced Aciplex® membranes, respectively.

Membranes can generally be categorised into three groups: (1) acid-based PEMs, such as phosphoric acid-doped Nafion®–PBI complex membranes; (2) modified PFSA membranes, which incorporate hydrophilic oxides and solid inorganic proton conductors; and (3) complex membranes, such as PVDF, PBI and SPSF, which are made from alternative sulfonated polymers. The development of chemically stable membrane against proxy radicals has gained significant awareness from researchers. One way to solve this is to develop new membranes with high chemical stability. For example, with low resistance to radicals, the radiation-grafted FEP-g-polystyrene membrane utilised as a sacrificial material [28–31]. During membrane fabrication, free radical inhibitors as well as stabilisers such as antioxidants or hindered amines can be mixed through the fabrication membrane. The chemical stability of the membrane was remarkably augmented by excluding the unstable end group [25, 32]. In addition to that, redesigning the MEA mitigates the chemical damage that can occur due to hydrogen peroxide. Nakagawa [33] proposed a complex membrane where a thin modified Nafion membrane was bonded with a membrane of PSSA. Via this technique, degradation of PSSA membrane cause by oxidation was successfully eliminated.

Moreover, the addition of peroxide-decomposition catalysts, such as heteropoly acids, into membranes has led to less membrane degradation. The use of new corrosion resistant metal coatings and catalysts that produce less hydrogen peroxide can massively reduce the degradation of membranes.

2.3.2 Degradation of Catalyst Layers and Electro-Catalysts

The electro-catalysts of PEM fuel cells are made of ternary, Pt and binary, quaternary Pt-transition metal alloys. Their conductive supports are usually made of high-surface-area carbon materials, such as Black Pearls BP2000, Ketjenblack or Vulcan XC 72. In theory, these catalysts can meet the cost requirements and performance expectations of high-volume applications. However, performance and reliability of currently used materials are still below expectations, especially, under severe operating environments, such as high humidity, dynamic loads and low pH values, in combination with high temperatures and oxidations.

2.3.2.1 Mechanisms of Catalyst Layer Degradation

Significant efforts have been made to understand the process of platinum (Pt) catalyst degradation caused by long-term operation, in particular, under harsh environments. Initially, a pure Pt catalyst might be exposed to impurities created by supply reactants [34, 35]. Moreover, there is a possibility that the catalyst may become non-active due to sintering of Pt particles, corrosion of the carbon support and detachment or dissolution of Pt into the membrane. The following factors have been identified as major contributors to coarsening of catalysts:

1. Small Pt particles may dissolve in the ionomer phase and deposit again on the surface of large particles, causing the particle to grow, a phenomenon known as Ostwald ripening [36]. The dissolved Pt species may diffuse into the ionomer phase and eventually precipitate in the membrane through reduction of Pt ions by the crossover hydrogen from the anode side, drastically reducing the membrane conductivity and stability.
2. Random cluster–cluster collisions, the agglomeration of platinum particles on the carbon support may take place at the nanometre scale causing a typical log-normal distribution of particle sizes with a maximum at smaller particle sizes and a tail towards the larger particle sizes.
3. Minimisation of the clusters (also known as Gibbs free energy). The catalyst particles may grow according to the atomic scale. Here, the distribution of the particle size can be characterised/identified by a tail where particle size is small and a maximum at the larger particle sizes [37]. However, there still no agreement on identifying the mechanism that governs and is principally responsible for the catalyst particle growth. Coalescence on the carbon support and coarsening of the catalyst because of movement of particles can cause the catalytically active surface area (ASE) to reduce. Lastly, there is probably an increase in particle sizes that leads to the reduction in catalyst activity due to the development of metal oxides at the anode or cathode side.

Another significant matter related to catalyst durability, which has been subject to considerable attention, is the catalyst carbon support corrosion [38]. Both modes are considered to induce carbon corrosion:

1. Starvation of fuel because of H_2 obstruction from a section of the anode under steady-state conditions
2. Transitioning between start-up and shutdown cycles. First mode, which is starvation of fuel in individual fuel cells, may arise from imbalanced flow sharing between one cell and other cells during high complete stack application or from gas flow obstruction ascribed to ice formation when fuel cells work in sub-zero temperatures. The second mode, denoted as the air-fuel front, can result from non-uniform allocation of fuel on the anode and the probable crossover of oxygen through the membrane during start-up and shutdown of the PEM fuel cells. In either of the cases, the anode electrode is moderately covered with hydrogen. Under the conditions of hydrogen exhaustion, the anode voltage is driven to negative until water and carbon oxidation occurs.

Despite the thermodynamic variability, carbon corrosion in a typical PEM fuel cell is insignificant at voltages less than 1.1 V compared to reversible hydrogen electrodes because of slow kinetics. Recent experiments have shown that the occurrence of Pt/C or PtRu/C electro-catalysts can contribute to hasten carbon corrosion and decrease the voltage for carbon oxidation to lower than, or equal to, 0.55 V (VS. RHE) [35, 37].

Carbon in general is protected from corrosion by the process of water oxidation when there is adequate water in the fuel cell, unless the water in the electrodes is exhausted, or the cell is exposed to a high current density not supportable by water oxidation alone. There is a potential that cell reversal, because of fuel starvation, can affect the durability of the GDL, the Bp or the CL. As an outcome, the comparative percentage of conductive material in the electrodes may decrease, while the internal resistance of the cell, and the contact resistance with the current collector, may eventually increase. Under acute conditions, the availability of the number of sites for anchoring the catalyst reduces with carbon corrosion, resulting in catalyst metal sintering [37], and in severe cases, a disintegration of the electrodes structure. At sub-zero temperatures, there is another significant hazard to PEM FC durability, which is the impact of volume changes of water and phase transformation on the physical characteristics of the PEM and on the electrode's interface and structure. It was observed by Schmittinger [38] that there is a decline in performance at a rate of about 2.3% per freeze–thaw cycle from (80 to -10) °C. The physical damage of the electrode's structure and MEA integrity caused by ice expansion during freezing is often considered to be a factor behind the degradation in cell performance observed in thermal cycles. Kusoglu et al. [39] presented analytical formulae that validate the impact of temperature cycling (between 80 and -40 °C) on the ionic conductivity, membrane structure, mechanical strength, water management and gas permeability. Reference [39] provides a comprehensive synopsis of the research on PEM fuel cells' freeze and rapid start-up. The results obtained from the experiments of Xie et al. [40] show the variation in hydrophobic properties of the CLs that takes place eventually because of the dissolution of PTFE or Nafion, which adversely impacts the mass transport potential of the and water management.

2.3.2.2 Catalyst Layer Degradation: Mitigation Techniques

Several methods have been proposed to augment the durability of catalysts. To begin with, operating environments have significant role in the process of catalyst degradation. The Pt catalysts are more stable at the anode side than those at the cathode are. Experimental results obtained by Zarrin et al. [41] indicate that the loss or reduction of Pt active surface areas could be remarkably reduced by operating at low relative temperature and humidity (RH). Borup et al. [42] recently discovered that the carbon corrosion of the catalysts increases with a decrease in RH. It was also found that the increase in cathode Pt particle size was significantly less during steady-state testing than during cycling experiments. The Pt particle size increased with an increase in voltage, which was validated during the recent application of an AST method to analyse the stability of the electro-catalyst. Then, by improving the water control on the anode, the corrosion of carbon support, as a result of fuel starvation, can be lessened through the addition of water-blocking components such as graphite, modifications to the ionomer and the use of improved catalysts for water electrolysis [43].

The first proposed solution to mitigate degradation of PEM fuel cells is called 'keep-warm' technique [44–46]. The technique is based on ensuring the temperature of the system is above a particular threshold. This is usually secured by consuming energy from the converter or an extra battery, especially, when the system is in a non-operational state. The second solution is to increase system's temperature by ensuring water temperature, inside the fuel cell, is well above freezing point at starting-up. As a higher energy temperature is required for this solution, it is highly recommended to combine it with an accurate procedure to remove residual water. This enables energy savings and stops any possible physical damage to MEA caused by ice expansion. Gas disinfection or washing with an antifreeze solution before stopping the fuel cells is some of the feasible techniques to get rid of the water. Third, compared to pure Pt catalysts, Pt–alloy catalysts, for example, Pt–Cr–Ni and Pt–Co, have exhibited good stability. The improvement may be attributed to an increased resistance due to the large size of the alloy particles or elements of the alloy. In spite of this, X-ray diffraction analysis has detected a skin that constitutes a monolayer of pure Pt formed on the surface of alloys after long-term testing [47, 48]. This implies that the nonnoble metals present in the Pt-transition metal alloy catalysts are more predisposed to being dissolved in the ionomer phase, and thus the benefit of Pt–alloy catalysts is partly neutralised. Cr, Co, Ni, V and Fe are identified as good materials, but Pt–Co/C has gained a lot of attention owing to its enhanced stability [49]. Considerable improvement, on the stability of Pt against dissolution under prospective cycling testing, has been achieved by altering its nanoparticles using gold (Au) clusters [50]. Under oxidising environments and prospective cycling between 0.6 and 1.1 V after over 30,000 cycles, no noticeable changes were observed in the activity and surface area of Au-altered Pt nanoparticles. The significant enhancement in the stability of the Au/Pt/C catalyst was primarily due to the occurrence of insoluble Au clusters. The sintering and dissolution potential of the metal alloy catalysts can be improved by reinforcing the interaction between the carbon support and the metal particles. Maiyalagan et al. [51] proposed and initiated a nitrogen-based carbon functionality to the carbon support surface through chemical alteration, and thus, a considerable improvement was noted in the ability of the treated support to anchor metal particles as well as its catalyst activity. In PEM fuel cell applications, multiwalled carbon nanotubes (CNTs) have also exhibited the potential of catalyst supports. Shao et al. [52] mentioned that under same conditions of an accelerated durability test, the rate of degradation of Pt/CNTs was almost two times lower than that of Pt/C. This phenomenon was ascribed to the distinctive interaction between CNTs and Pt and to the increased resistance of the CNTs to the oxidation reactions. Moreover, the reduction in support surface area or graphitisation of the carbon support can increase the support's resistance to oxidation and carbon corrosion. Despite of this, the number of active surface sites that anchor the metal particles correspondingly reduces, which is possibly detrimental to the accumulation of metal on the carbon support.

2.3.3 Gas Diffusion Layers

2.3.3.1 Degradation Mechanism of Gas Diffusion Layers

In general, the gas diffusion layers (GDLs) are a carbon-based dual-layer made up of porous material, which consists of carbon cloth substrate or microporous carbon fibre paper protected by a thinner microporous layer consisting of a hydrophobic agent and carbon black powder. In earlier research of GDLs, the focus was on the effect of GDL design and materials on the performance PEM fuel cells, instead of durability. Nevertheless, after 11,000 h of operation and cold start conditions [53], augmented GDL surface hydrophilicity has evidently been observed and implies with certainty that further investigation of the GDL is necessary. So far, only a few studies have concentrated on the correlation between GDL properties and performance or the degradation mechanisms of GDLs. To sum up, these studies have primarily used ex situ GDL ageing methodologies to escape the potential confounding effects from the adjacent elements such as the bipolar plate (Bp) and the CL. Yang et al. [54] indicate that the loss of GDL hydrophobicity improved with operating temperature and with the use of air sparging in place of nitrogen. Moreover, the conclusion was reached that alterations in the GDL characteristics were ascribed typically to the microporous layer. Tanuma et al. [55] made the GDLs grow by immersing the samples in 15wt percentage hydrogen peroxide at 82 °C. With this experiment, it was discovered that the greater the time of exposure, the greater was the weight loss and the MPL contact angle, and the increase was accredited to oxidation of the carbon in the microporous layer. Kangasniemi et al. [56] explained the impact of electrochemical surface oxidation on the GDL characteristics and discovered that the contact angle of the microporous layer surface reduced significantly with time when the GDL samples were submerged in 1 M H_2SO_4 under potentiostatic treatment of 1.2 V vs. standard hydrogen electrode (SHE). In recent times, Oberholzer et al. [57] investigates the effects of bending stiffness, electrical resistivity, surface contact angle, air permeability, water vapour diffusion and porosity on the GDL characteristics. Tested cases included after degradation tests at steady-state conditions (over 1500 having time at 80 °C and 13.6 atm) and freezing (54 freeze–thaw cycles between −35 and 20 °C) conditions. The carbon composite and PTFE of the GDLs are predisposed to electrochemical (voltage) oxidation and chemical attack (i.e. OH• radical as electrochemical by-product), during the FC operation. With the loss of carbon and PTFE, alterations are observed in the physical characteristics of GDL changes, such as there was a reduction in the GDL conductivity and hydrophobicity, which further decreased the

MEA performance and negatively impacted the durability of the complete (or overall) FC. The relationship between changes in GDL characteristics and performance loss was studied in Ref. [58].

2.3.3.2 Mitigation Techniques for Gas Diffusion Layer Degradation

There is not much literature about methods to eliminate the degradation of GDL. However, to enhance the electro-oxidative stability and GDL oxidative, Borup [42] proposed the use of graphitised fibres during GDL preparation. Borup further suggested that higher PTFE loading may prove to be advantageous to the water management potential of aged GDLs. Ha et al. [59] observed a 25% increase in the start/stop degradation rate (at 1.2 A/cm^2) caused by the use of graphitised carbon material Pureblack® in the MPL.

2.3.4 Bipolar Plate

2.3.4.1 Degradation Mechanisms of Bipolar Plate

The bipolar plate (Bp) that functions as a separator between the coolant hydrogen and air is a multifunctional component of the fuel cell system, which consistently distributes the reactants and produces streams, providing a path to current that produces during the electrochemical reactions. A significant amount of research has been made on the Bp component, and several literature reviews [60–62] have also been published. Several characteristics are required for the materials to be suitable for Bp component so that all the necessary functions are accomplished. These characteristics include, but are not restricted to, low gas permeability, high ionic and protonic conductivity, sufficient strength, low thermal resistance, high resistance to corrosion and reasonable manufacturing costs. At present, many materials, such as polymer-based composites with conductive carbon/graphite fillers and graphite, are being evaluated [63].

The combination of high thermal and electrical conductivity, high current density and high corrosion and chemical resistance is advantageous properties of graphite and graphite composites. However, their suitability to hydrogen, manufacturability, durability under shock and vibration is low in comparison to metals. To overcome these inadequacies, the volume and weight of electrodes should be increased. Common metals, e.g. Bps, are Nb, Pt, Zr and Ta, are extremely resistant to corrosion and can be produced in light weight thin Bps; however, the raw material for these Bps limits their use in commercial applications. Commercially obtainable metals and their alloys such as Ni, Al and Ti have shown strong potential for their use in Bps due to their good mechanical characteristics, high electrical conductivity and low cost. Nevertheless, these metallic Bps pose a significant challenge because of their contact resistance between the Bp and the GDL, which is ascribed to the formation of electrically resistant oxide films on the surfaces that unavoidably increase the internal electrical resistance. For researchers, stainless steel has been the focus of significant attention because of its wide range of alloys and applicability to mass production. However, stainless steel is susceptible to corrosion in aggressive humid and acidic environments, in particular, that inside a PEM fuel cells [64], which also increases the contact resistance. In addition, metallic materials undergo corrosion to produce multivalent cations, which may adversely affect the membrane and catalyst durability [54, 65]. After 100 h of operation, the MEA chemical analysis showed that untreated stainless steel 316 L discharged large amounts of Ni and Fe atoms and traces of Cr [65]. As a result, the corrosion led to an electrical potential drop of up to 300 mV at a measured current of 700 mA/cm^2. Brady et al. [66] conducted an ex situ experiment to analyse the chemical corrosion in eight commercial stainless steel bipolar plates exposed to acidic solution. A controlled voltage was applied, and the solution was bubbled with a gas (H_2 or air) – to simulate either cathodic or anodic conditions. After 72 h, the solution was found to contain dissolved metal cations (including Fe^{+3} and Cr^{+3} Ni^{+2}).

2.3.4.2 Mitigation Techniques for Bipolar Plate Degradation

Recently reported methods to lessen degradation of bipolar plates proposed the use of graphite/polymer composites, carbide-based alloys or various nitride or metallic materials coated with noble metals [67]. A number of developed coating methods, widely employed in other industrial areas, for example, electro less deposition, spraying, electroplating, immersion coating, painting and electrolytic anodization, were examined for Bp materials [68]. Considering manufacturing cost-effectiveness for continuous high-volume production, chemical vapour deposition (CVD) and physical vapour deposition (PVD) processes has been reported [69]. However, the protective coating methods may face potential risk of coating material deformation during the operation of PEM fuel cell under high thermal conditions.

Kakati et al. [68] suggested a sensible description that the difference in the coefficients of thermal expansion (CTE) of coating material and the substrate causes their expansion and contraction at various rates. The deformation of the coating layer may cause microcracks and micropores, which exposes the substrate metal directly to an extremely corrosive environment. This subsequently leads to the diffusion of dissolved metal ions into the PEM, thus being trapped in the ion exchange sites, which adversely affects the performance of the FC. One effective strategy to buffer CTE differential is to add intermediate coating layers with high bonding strength and introduce gradient CTEs between coating and adjacent layers of the Bp.

The compressive forces employed to guarantee electric contact between the layers and reactant sealing when the cell in operating state may cause deformation or even fracture, which is a common concern related to Bps. In addition, operational factors such as non-uniform current, thermal cycling or thermal misdistributions over the active area can affect the bipolar plate's mechanical properties. Lee et al. [69], in the presence of an acidic electrolyte, introduced electrical current to the surface of a metallic material (nickel-rich alloys or stainless steel). This treatment enhanced the corrosion resistance property of the metallic Bp due to the surface composition and/or the modification of the surface morphology. Delamination is not an issue here as the surface treatment method used here modifies only the surface rather than the coating procedure.

Although carbon–carbon composites lack the mechanical strength, as well as needing the lengthy and costly process of chemical vapour impregnation (CVI), they offer many advantages. In recent times, carbon–polymer materials, particularly those with thermoset resins, are quickly gaining popularity as alternatives for Bps in terms of dimensional tolerance and good conductivity. However, the required electrical conductivity requires high carbon loadings, which may cause difficulties in the ability to process. Degradation of the resins in PEM fuel cell working environments owing to the polymer's inherent properties is another issue related to the materials of carbon–polymer. The heavy atoms, such as magnesium, calcium or zinc, released from these resins may diffuse into and contaminate the membrane, which reduces the performance and durability of fuel cells.

While no material meets all of the desirable properties that suit industrial applications, the selection criteria is, in most cases, a compromise between reliability, durability, efficiency, performance, lifetime, size, cost, local climate and operational range flexibility.

2.4 Monitoring Techniques for PEM Fuel Cells

A PEM fuel cell stack is a complicated system that consists of several components and thus establishing effective monitoring tool requires in-depth understanding of component interactions, degradation mechanisms, material science and the effects of operating conditions.

Many approaches have been implemented by several developers to evaluate the performance end condition of PEM fuel cells. Efforts have also been made to evaluate and improve their reliability, durability as well as failure modes to reduce experimental time and improve the accuracy. Currently, the US FC Council (USFCC) and the DOE are putting efforts into establishing PEM fuel cell durability and reliability testing protocols and set standards for operating procedures and test conditions to examine new structures and component materials.

Typical tests made through this suite include electrochemical, electrical and mechanical tests, and associated measurements are performed according to two modes: fuel cell in operation and in ex situ measurements.

The most common ex situ tests for the operating fuel cells are described in this section together with reference to applications to contamination testing.

2.4.1 Polarisation Curve

The polarisation curve (also known as the V-I curve) is one of the most widely used performance testing approaches. This curve consists of several recordings of cell voltages as a response to current or current density levels. When the current increases from zero for an open circuit, the electrodes begin to polarise, producing lower cell voltage. This curve includes essential operating characteristics for various applications, since it determines the available power. The curve's shape provides vital information about the magnitude of the polarisation losses, categorised as kinetic, ohmic and mass transport (see Chap. 3) [2].

2.4.2 Steady-State Testing

This is the second common test, and it is used to investigate the durability of the fuel cell. Such a type of test is quite simple and easy to perform, as it only requires setting a constant load or current, or in some cases, a constant voltage for the fuel cell under testing. Continue to operate for a specific period or until a failure criterion is reached, while changes in voltage and/or current are recorded. The rate and type, performance and degradation can be characterised by conducting diagnostics schemes, which can be made through polarisation curves, or some of other testes described in the following sections.

Fuel cell runs for a short period to produce a baseline performance and the contaminants are used into one of the inlet reactant streams to enable the monitoring of performance degradation. The contaminant may then be stopped, and the test continues to determine the possibility of having spontaneous recovery of any performance losses or that there is a need to conduct further tests, e.g. to detect the presence of adsorbed species on the catalyst [71].

2.4.3 Duty Cycle Testing

Based on application in question, fuel cell's simulation may be quite a complex process. As an example, on the efforts involved, and a study that was conducted by Sandia National Laboratories robotic vehicle [72], this study shows the effort made in characterising the duty cycle of the vehicle with a data acquisition system. Despite being a valuable tool, duty cycle testing is rarely used in the literature since it requires an automated test station and good understanding of the intended duty cycle.

2.4.4 Oxygen Testing

The oxygen polarisation curve provides a measure of the activation and ohmic losses, with minimal contribution from mass transport or concentration losses, as the diffusion rate of the oxygen is much higher than that of the binary gas diffusion with air. This justifies the need to separate the kinetic, ohmic and mass transport losses.

2.4.5 Voltage Decay

The voltage diagnostic technique is based on the analysis of the transient voltage response after step change in load from zero to a constant current. The test is performed while running the cell under the conditions of interest until it becomes stable, and the cell is then set to 'open circuit'. The current value increased from zero to the specific value depending on the load, and the transient voltage response is measured.

The initial response occurs because of reactants are at a uniform concentration as there is no current being generated. Just after the load is applied, the voltage (V_o) rises before a diffusion gradient can be established in the electrodes, while the reactant concentration at the CL is the same as the bulk gas concentration. It is worth mentioning that with a uniform distribution of temperature and reactant concentration, V_o should occur with a uniform current distribution. Repeating this test (via stepping to different load points) can produce a mass transport free polarisation curve. As reactants are being consumed, a diffusion gradient is established, and the subsequent response provides information about mass transport.

2.4.6 Alternating Current Impedance Testing

The alternating current (AC) impedance testing is different from the majority of other diagnostic techniques as it applies a very small perturbation to the system in the form of an alternating signal. The small signal results in reduced errors introduced by the measurement technique. The theory employs the equivalent electronic circuit consisting of resistors, capacitor and inductors to represent the impedance to electron flow in an electrochemical cell, such as the slow electrode kinetics and diffusion resistance. The theory uses an equivalent electronic circuit consisting of resistors, a capacitor and inductors to represent the impedance to electron flow in an electrochemical cell, such as the slow electrode kinetics and diffusion resistance. Relatively simple equivalent circuit models are frequently good approximations of a real system, and data can be fitted to yield reasonably accurate results. By introducing a perturbation in the direct current (DC) delivered by fuel cell and measuring both the voltage response and the perturbed current, an AC impedance can be calculated. Electrochemical impedance spectroscopy (EIS) involves the measurement of the response at several frequencies.

Vengatesan et al. [73] provided an example of the use of AC impedance for assessing the degradation over a lifetime test. When measuring the AC impedance spectra at different current densities for new and lifetime-tested cells, the resistance to transport, or diffusion processes, has significantly increased between the new cell and the life-tested cell. By applying the proposed technique under different humidification levels, the authors were able to confirm that there is a connection between the presence of water and the increased resistance to diffusion. The study also considered the change in ionic conductivity of the electrodes.

2.4.7 Cyclic Voltammetry Testing

This type of tests is commonly used to investigate electrodes and hydrogen crossover. In this technique, linear 'sweep' potential is applied to one electrode, while the other is held constant. The potential is cycled in a 'triangular wave' pattern, and the current produced is monitored.

The information expressed by the shape and magnitude of the current response are useful quantitative and qualitative information regarding the amount of catalyst, that is, electrochemically active, the double layer capacitance, hydrogen crossover, and the presence of oxide layers and contaminants. Wu et al. [70] provide a description of this technique with example on voltammograms.

2.4.8 Single Cell Versus Stack Testing

PEM fuel cells developers commonly use a mixture of single test and stack testing. Both types of testing have several advantages as follows:

The main advantages of single testing are (1) it is cost effective test, (2) it provides great control of operating conditions and (3) it is more useful for fundamental studies.

The main advantages of stack testing are: the ability to simulate very closely operation under realistic conditions including the range of conditions experienced by the cells in the stack.

The main differences for operation in a stack are due to uneven thermal, electrical and flow distribution effects between the cells. Moreover, the heat loss from a single cell is quite large, and this makes it quite difficult to simulate the same thermal conditions as a cell would see in a stack. This can lead to some differences for contamination studies, and it is well-known that temperature effects are very important for adsorption and desorption phenomena.

2.4.9 Accelerated Durability Testing

To make testing of fuel cells more realistic, long-life requirement should be considered in the test. Applications such as stationary power generation at 40,000 h and bus application at 20,000 h will take several years to test and will need significant quantities of fuel. The development and design cycle require turnaround in a make–break–fix cycle involving determination of failure modes, followed by technology development, and, finally, implementation of mitigating materials and designs. Therefore, the main features are if the FC testing approach is sufficient for the required lifetime; thus accelerated testing techniques are being developed by many FC developers and material suppliers and are an important component of fuel cell testing strategy.

The use of an accelerated stress testing method not only stimulates the specific component's targeted failure mode but also minimises the confounding effects from other components. As the components undergo different degradation mechanisms under various conditions, AST protocol for catalyst supports differ when compared with electro-catalysts. Similarly, AST requires isolating and distinguishing of the effects of membrane chemical degradation for understanding membrane mechanical degradation.

2.5 Summary

By reviewing the abovementioned research effort, it was noted that performance, reliability and durability are three major issues which are obstacles to the commercialisation of fuel cells.

The performance, efficiency and reliability of PEM fuel cells are affected by many internal and external factors, such as the degradation of materials, operational conditions and impurities or poisons. Undoubtedly, while more conventional experimental approaches dedicated to cell state assessment (electrochemical impedance spectroscope, current interruption, pressure drop and X-ray method) are undoubtedly powerful, they remain unsuitable in terms of handling the heterogeneous, transient and even stochastic elements of the phenomena witnessed in the membrane of the fuel cells.

Phenomena that occur within PEM fuel cells are many, and sufficient understanding of these shall lead to well understanding of failures occurring in fuel cell systems. This will help to establish the diagnosis and monitoring systems are of high reliability and ultimately enhance the reliability of the cell and reduce its cost. Essentially, due to the cell geometry, chemical, electrical and physical values, uniformity is not flawlessly achieved. In some regards, an approach facilitating the membrane hydrogen state measurement which needs to be sensitive to localised phenomena would significantly contribute to developments made in water management.

Although all methods used to for condition monitoring and diagnosing failures are generally well understood and give acceptable results, using them with stack fuel cell is a difficult task, and they require more complex procedures in order to obtain good-quality data. Another issue is the cost, where all techniques utilising monitoring and diagnosis faults are expensive.

References

1. Fecarotti, C., Andrews, J., & Chen, R. (2016). A Petri net approach for performance modelling of polymer electrolyte membrane fuel cell systems. *International Journal of Hydrogen Energy, 41*, 12242–12260.
2. Alrweq, M., & Albarbar, A. (2016). Investigation into the characteristics of proton exchange membrane fuel cell-based power system. *IET Science, Measurement & Technology*. doi: https://doi.org/10.1049/iet-smt.2015.0046, Online ISSN 1751–8830.
3. James, B. D., & Spisak, A. B. (2012). *Mass Production Cost Estimation of Direct H2 PEM Fuel Cell Systems for Transportation Applications: 2012 Update*. Report by Strategic Analysis, Inc., under Award Number DEEE0005236 for the US Department of Energy, 18.and sustainable energy reviews, *11*(8), 1720–1738.
4. Zhou, Y., Neyerlin, K., Olson, T. S., Pylypenko, S., Bult, J., Dinh, H. N., Gennett, T., Shao, Z., & O'Hayre, R. (2010). Enhancement of Pt and Pt-alloy fuel cell catalyst activity and durability via nitrogen-modified carbon supports. *Energy & Environmental Science, 3*(10), 1437–1446.
5. Devanathan, R. (2013). Proton exchange membranes for FCs. In R. H. Crabtree (Ed.), *Energy production and storage: Inorganic chemical strategies for a warming world* (Vol. 8, p. 89.) ISBN: 978-0-470-74986-9.
6. Scott, K., Xu, C., & Wu, X. (2014). Intermediate temperature proton-conducting membrane electrolytes for fuel cells. *Wiley Interdisciplinary Reviews: Energy and Environment, 3*(1), 24–41.
7. Subianto, S., Pica, M., Casciola, M., Cojocaru, P., Merlo, L., Hards, G., & Jones, D. J. (2013). Physical and chemical modification routes leading to improved mechanical properties of perfluorosulfonic acid membranes for PEM fuel cells. *Journal of Power Sources, 233*, 216–230.
8. Tang, H., Qi, Z., Ramani, M., & Elter, J. F. (2006). PEM fuel cell cathode carbon corrosion due to the formation of air/fuel boundary at the anode. *Journal of Power Sources, 158*(2), 1306–1312.
9. Cheng, X., Zhang, J., Tang, Y., Song, C., Shen, J., Song, D., & Zhang, J. (2007). Hydrogen crossover in high-temperature PEM fuel cells. *Journal of Power Sources, 167*(1), 25–31.
10. Gittleman, C. S., Coms, F. D., & Lai, Y. H. (2011). Membrane durability: Physical and chemical degradation. In *Polymer electrolyte fuel cell degradation* (pp. 15–88). Denmark: Elsevier.
11. Gubler, L., & Scherer, G. G. (2009). Durability of radiation-grafted fuel cell membranes. In *Polymer electrolyte fuel cell durability* (pp. 133–155). New York: Springer.
12. Young, A. P. (2010). *Characterization of structural degradation in a polymer electrolyte membrane fuel cell cathode catalyst layer* (Doctoral dissertation, University of British Columbia).
13. Wang, S., Jiang, S. P., White, T. J., Guo, J., & Wang, X. (2009). Electrocatalytic activity and interconnectivity of Pt nanoparticles on multi-walled carbon nanotubes for FCs. *The Journal of Physical Chemistry C, 113*(43), 18935–18945.
14. Carmo, M., Fritz, D. L., Mergel, J., & Stolten, D. (2013). A comprehensive review on PEM water electrolysis. *International Journal of Hydrogen Energy, 38*(12), 4901–4934.

15. Zhou, X., Qiao, J., Yang, L., & Zhang, J. (2014). A review of graphene-based nanostructural materials for both catalyst supports and metal-free catalysts in PEM fuel cell oxygen reduction reactions. *Advanced Energy Materials, 4*(8), 1301523.
16. Tang, H., Peikang, S., Jiang, S. P., Wang, F., & Pan, M. (2007). A degradation study of Nafion proton exchange membrane of PEM fuel cells. *Journal of Power Sources, 170*(1), 85–92.
17. Chandan, A., Hattenberger, M., El-Kharouf, A., Du, S., Dhir, A., Self, V., Pollet, B. G., Ingram, A., & Bujalski, W. (2013). High temperature (HT) polymer electrolyte membrane fuel cells (PEMFC)–A review. *Journal of Power Sources, 231*, 264–278.
18. Subban, C. V., Zhou, Q., Hu, A., Moylan, T. E., Wagner, F. T., & DiSalvo, F. J. (2010). Sol–gel synthesis, electrochemical characterization, and stability testing of Ti0. 7W0. 3O2 nanoparticles for catalyst support applications in proton-exchange membrane fuel cells. *Journal of the American Chemical Society, 132*(49), 17531–17536.
19. Wu, J., Yuan, X. Z., Martin, J. J., Wang, H., Zhang, J., Shen, J., Wu, S., & Merida, W. (2008). A review of PEM fuel cell durability: Degradation mechanisms and mitigation strategies. *Journal of Power Sources, 184*(1), 104–119.
20. Kraytsberg, A., & Ein-Eli, Y. (2014). Review of advanced materials for proton exchange membrane fuel cells. *Energy & Fuels, 28*(12), 7303–7330.
21. Abdo, N., & Easton, E. B. (2016). Nafion/Polyaniline composite membranes for hydrogen production in the Cu–Cl thermochemical cycle. *International Journal of Hydrogen Energy, 41* (19), 7892–7903.
22. Okada, T., Møller-Holst, S., Gorseth, O., & Kjelstrup, S. (1998). Transport and equilibrium properties of Nafion® membranes with H+ and Na+ ions. *Journal of Electroanalytical Chemistry, 442*(1), 137–145.
23. Gancs, L., Hult, B. N., Hakim, N., & Mukerjee, S. (2007). The impact of Ru contamination of a Pt/C electrocatalyst on its oxygen-reducing activity. *Electrochemical and Solid-State Letters, 10*(9), B150–B154.
24. Inaba, M., Kinumoto, T., Kiriake, M., Umebayashi, R., Tasaka, A., & Ogumi, Z. (2006). Gas crossover and membrane degradation in polymer electrolyte fuel cells. *Electrochimica Acta, 51*(26), 5746–5753.
25. Rowshanzamir, S., Peighambardoust, S. J., Parnian, M. J., Amirkhanlou, G. R., & Rahnavard, A. (2015). Effect of Pt-Cs 2.5 H 0.5 PW 12 O 40 catalyst addition on durability of self-humidifying nanocomposite membranes based on sulfonated poly (ether ether ketone) for proton exchange membrane fuel cell applications. *International Journal of Hydrogen Energy, 40*(1), 549–560.
26. Jeon, Y., Park, J. I., Ok, J., Dorjgotov, A., Kim, H. J., Kim, H., Lee, C., Park, S., & Shul, Y. G. (2016). Enhancement of catalytic durability through nitrogen-doping treatment on the CNF-derivatized ACF support for high temperature PEMFC. *International Journal of Hydrogen Energy, 41*(16), 6864–6876.
27. Koh, S., Leisch, J., Toney, M. F., & Strasser, P. (2007). Structure-activity-stability relationships of Pt-Co alloy electrocatalysts in gas-diffusion electrode layers. *The Journal of Physical Chemistry C, 111*(9), 3744–3752.
28. Dubau, L., Castanheira, L., Maillard, F., Chatenet, M., Lottin, O., Maranzana, G., Dillet, J., Lamibrac, A., Perrin, J. C., Moukheiber, E., & ElKaddouri, A. (2014). A review of PEM fuel cell durability: Materials degradation, local heterogeneities of aging and possible mitigation strategies. *Wiley Interdisciplinary Reviews: Energy and Environment, 3*(6), 540–560.
29. Vielstich, W., Yokokawa, H., & Gasteiger, H. A. (2009). *Handbook of fuel cells: Fundamentals technology and applications*. Chichester: Wiley.
30. Zhang, H., & Shen, P. K. (2012). Recent development of polymer electrolyte membranes for fuel cells. *Chemical Reviews, 112*(5), 2780–2832.
31. Jeon, Y., Na, H., Hwang, H., Park, J., Hwang, H., & Shul, Y. G. (2015). Accelerated life-time test protocols for polymer electrolyte membrane fuel cells operated at high temperature. *International Journal of Hydrogen Energy, 40*(7), 3057–3067.

32. Peighambardoust, S. J., Rowshanzamir, S., & Amjadi, M. (2010). Review of the proton exchange membranes for fuel cell applications. *International Journal of Hydrogen Energy, 35*(17), 9349–9384.
33. Nakagawa, T., Nakabayashi, K., Higashihara, T., & Ueda, M. (2010). A high performance polymer electrolyte membrane based on sulfonated poly (ether sulfone) with binaphthyl units. *Journal of Materials Chemistry, 20*(32), 6662–6667.
34. Xin, H. L., Mundy, J. A., Liu, Z., Cabezas, R., Hovden, R., Kourkoutis, L. F., Zhang, J., Subramanian, N. P., Makharia, R., Wagner, F. T., & Muller, D. A. (2011). Atomic-resolution spectroscopic imaging of ensembles of nano-catalyst particles across the life of a fuel cell. *Nano Letters, 12*(1), 490–497.
35. Meier, J. C., Galeano, C., Katsounaros, I., Topalov, A. A., Kostka, A., Schüth, F., & Mayrhofer, K. J. (2012). Degradation mechanisms of Pt/C fuel cell catalysts under simulated start–stop conditions. *ACS Catalysis, 2*(5), 832–843.
36. Yuan, X. Z., Li, H., Zhang, S., Martin, J., & Wang, H. (2011). A review of polymer electrolyte membrane fuel cell durability test protocols. *Journal of Power Sources, 196*(22), 9107–9116.
37. Pedersen, C. M., Escudero-Escribano, M., Velázquez-Palenzuela, A., Christensen, L. H., Chorkendorff, I., & Stephens, I. E. (2015). Benchmarking Pt-based electrocatalysts for low temperature fuel cell reactions with the rotating disk electrode: Oxygen reduction and hydrogen oxidation in the presence of CO (review article). *Electrochimica Acta, 179*, 647–657.
38. Schmittinger, W., & Vahidi, A. (2008). A review of the main parameters influencing long-term performance and durability of PEM fuel cells. *Journal of Power Sources, 180*(1), 1–14.
39. Kusoglu, A., Karlsson, A. M., Santare, M. H., Cleghorn, S., & Johnson, W. B. (2006). Mechanical response of fuel cell membranes subjected to a hygro-thermal cycle. *Journal of Power Sources, 161*(2), 987–996.
40. Xie, J., Wood, D. L., Wayne, D. M., Zawodzinski, T. A., Atanassov, P., & Borup, R. L. (2005). Durability of PEFCs at high humidity conditions. *Journal of the Electrochemical Society, 152* (1), A104–A113.
41. Zarrin, H., Higgins, D., Jun, Y., Chen, Z., & Fowler, M. (2011). Functionalized graphene oxide nanocomposite membrane for low humidity and high temperature proton exchange membrane fuel cells. *The Journal of Physical Chemistry C, 115*(42), 20774–20781.
42. Borup, R., Meyers, J., Pivovar, B., Kim, Y. S., Mukundan, R., Garland, N., Myers, D., Wilson, M., Garzon, F., Wood, D., & Zelenay, P. (2007). Scientific aspects of polymer electrolyte fuel cell durability and degradation. *Chemical Reviews, 107*(10), 3904–3951.
43. Lohoff, A. S., Günther, D., Hehemann, M., Müller, M., & Stolten, D. (2016). Extending the lifetime of direct methanol fuel cell systems to more than 20,000 h by applying ion exchange resins. *International Journal of Hydrogen Energy, 41*(34), 15325–15334.
44. Assarabowski, R.J., Unkert, W.T., Bach, L.A., Grasso, A.P., Olsommer, B.C., & Utc Fuel Cells, Llc (2004). Method and apparatus for preventing water in fuel cell power plants from freezing during storage. U.S. *Patent 6,797,421.*
45. Alink, R., Gerteisen, D., & Oszcipok, M. (2008). Degradation effects in polymer electrolyte membrane fuel cell stacks by sub-zero operation—An in situ and ex situ analysis. *Journal of Power Sources, 182*(1), 175–187.
46. Jung, H. M., & Um, S. (2011). An experimental feasibility study of vanadium oxide films on metallic bipolar plates for the cold start enhancement of fuel cell vehicles. *International Journal of Hydrogen Energy, 36*(24), 15826–15837.
47. Shao, M. H., Huang, T., Liu, P., Zhang, J., Sasaki, K., Vukmirovic, M. B., & Adzic, R. R. (2006). Palladium monolayer and palladium alloy electrocatalysts for oxygen reduction. *Langmuir, 22*(25), 10409–10415.
48. Shao, M., Shoemaker, K., Peles, A., Kaneko, K., & Protsailo, L. (2010). Pt monolayer on porous Pd– Cu alloys as oxygen reduction electrocatalysts. *Journal of the American Chemical Society, 132*(27), 9253–9255.

49. Gasteiger, H. A., Kocha, S. S., Sompalli, B., & Wagner, F. T. (2005). Activity benchmarks and requirements for Pt, Pt-alloy, and non-Pt oxygen reduction catalysts for PEMFCs. *Applied Catalysis B: Environmental, 56*(1), 9–35.
50. Sasaki, K., Shao, M., & Adzic, R. (2009). Dissolution and stabilization of platinum in oxygen cathodes. In F. N. Büchi, M. Inaba, & T. J. Schmidt (Eds.), *Polymer electrolyte fuel cell durability*. New York: Springer. https://doi.org/10.1007/978-0-387-85536-3_2.
51. Maiyalagan, T. (2008). Synthesis and electro-catalytic activity of methanol oxidation on nitrogen containing carbon nanotubes supported Pt electrodes. *Applied Catalysis B: Environmental, 80*(3), 286–295.
52. Shao, Y., Yin, G., Gao, Y., & Shi, P. (2006). Durability study of Pt/C and Pt/CNTs catalysts under simulated PEM fuel cell conditions. *Journal of the Electrochemical Society, 153*(6), A1093–A1097.
53. Meng, H. (2008). Numerical analyses of non-isothermal self-start behaviour of PEM fuel cells from subfreezing start-up temperatures. *International Journal of Hydrogen Energy, 33*(20), 5738–5747.
54. Yang, L., Larouche, N., Chenitz, R., Zhang, G., Lefèvre, M., & Dodelet, J. P. (2015). Activity, performance, and durability for the reduction of oxygen in PEM fuel cells, of Fe/N/C electrocatalysts obtained from the pyrolysis of metal-organic-framework and iron porphyrin precursors. *Electrochimica Acta, 159*, 184–197.
55. Tanuma, T., & Kinoshita, S. (2011). Impact of gas diffusion layers (GDLs) on water transport in PEFCs. *Journal of the Electrochemical Society, 159*(2), B150–B154.
56. Kangasniemi, K. H., Condit, D. A., & Jarvi, T. D. (2004). Characterization of Vulcan electrochemically oxidized under simulated PEM fuel cell conditions. *Journal of the Electrochemical Society, 151*(4), E125–E132.
57. Oberholzer, P., Boillat, P., Siegrist, R., Perego, R., Kästner, A., Lehmann, E., Scherer, G. G., & Wokaun, A. (2011). Cold-start of a PEFC visualized with high resolution dynamic in-plane neutron imaging. *Journal of the Electrochemical Society, 159*(2), B235–B245.
58. Lim, C., Ghassemzadeh, L., Van Hove, F., Lauritzen, M., Kolodziej, J., Wang, G. G., Holdcroft, S., & Kjeang, E. (2014). Membrane degradation during combined chemical and mechanical accelerated stress testing of polymer electrolyte fuel cells. *Journal of Power Sources, 257*, 102–110.
59. Ha, T., Cho, J., Park, J., Min, K., Kim, H. S., Lee, E., & Jyoung, J. Y. (2011). Experimental study on carbon corrosion of the gas diffusion layer in polymer electrolyte membrane fuel cells. *International Journal of Hydrogen Energy, 36*(19), 12436–12443.
60. Taherian, R. (2014). A review of composite and metallic bipolar plates in proton exchange membrane fuel cell: Materials, fabrication, and material selection. *Journal of Power Sources, 265*, 370–390.
61. Karimi, S., Fraser, N., Roberts, B., & Foulkes, F. R. (2012). A review of metallic bipolar plates for proton exchange membrane fuel cells: Materials and fabrication methods. *Advances in Materials Science and Engineering, 2012*, 1–22.
62. Iranzo, A., Muñoz, M., Rosa, F., & Pino, J. (2010). Numerical model for the performance prediction of a PEM fuel cell. Model results and experimental validation. *International Journal of Hydrogen Energy, 35*(20), 11533–11550.
63. Hu, Q., Zhang, D., Fu, H., & Huang, K. (2014). Investigation of stamping process of metallic bipolar plates in PEM fuel cell—Numerical simulation and experiments. *International Journal of Hydrogen Energy, 39*(25), 13770–13776.
64. Antunes, R. A., Oliveira, M. C. L., Ett, G., & Ett, V. (2010). Corrosion of metal bipolar plates for PEM fuel cells: A review. *International Journal of Hydrogen Energy, 35*(8), 3632–3647.
65. Gabreab, E. M., Hinds, G., Fearn, S., Hodgson, D., Millichamp, J., Shearing, P. R., & Brett, D. J. (2014). An electrochemical treatment to improve corrosion and contact resistance of stainless steel bipolar plates used in polymer electrolyte fuel cells. *Journal of Power Sources, 245*, 1014–1026.

66. Brady, M. P., Elhamid, M. A., Dadheech, G., Bradley, J., Toops, T. J., Meyer, H. M., & Tortorelli, P. F. (2013). Manufacturing and performance assessment of stamped, laser welded, and nitrided FeCrV stainless steel bipolar plates for proton exchange membrane fuel cells. *International Journal of Hydrogen Energy, 38*(11), 4734–4739.
67. Wang, Y. J., Qiao, J., Baker, R., & Zhang, J. (2013). Alkaline polymer electrolyte membranes for fuel cell applications. *Chemical Society Reviews, 42*(13), 5768–5787.
68. Kakati, B. K., Ghosh, A., & Verma, A. (2013). Efficient composite bipolar plate reinforced with carbon fiber and graphene for proton exchange membrane fuel cell. *International Journal of Hydrogen Energy, 38*(22), 9362–9369.
69. Lee, K., Zhang, J., Wang, H., & Wilkinson, D. P. (2006). Progress in the synthesis of carbon nanotube-and nanofiber-supported Pt electrocatalysts for PEM fuel cell catalysis. *Journal of Applied Electrochemistry, 36*(5), 507–522.
70. Wu, B., Lin, G., Fu, Y., Hou, M., & Yi, B. (2010). Chromium-containing carbon film on stainless steel as bipolar plates for proton exchange membrane fuel cells. *International Journal of Hydrogen Energy, 35*(24), 13255–13261.
71. Ni, M. (2013). Modeling and parametric simulations of solid oxide fuel cells with methane carbon dioxide reforming. *Energy Conversion and Management, 70*, 116–129.
72. Najafi, B., Mamaghani, A. H., Rinaldi, F., & Casalegno, A. (2015). Long-term performance analysis of an HT-PEM fuel cell based micro-CHP system: Operational strategies. *Applied Energy, 147*, 582–592.
73. Vengatesan, S., Panha, K., Fowler, M. W., Yuan, X. Z., & Wang, H. (2012). Membrane electrode assembly degradation under idle conditions via unsymmetrical reactant relative humidity cycling. *Journal of Power Sources, 207*, 101–110.

Chapter 3
Design and Fundamental Characteristics of PEM Fuel Cells

3.1 Overview

A fuel cell (FC) is a device that is capable of converting chemical energy into electrical energy and generating a direct electrical current ($I_{d.\ c}$) with weak direct voltage ($V_{d.\ c}$), gathered in a pack (i.e. a FC stack). Humidified hydrogen or other hydrocarbon fuel, such as methanol, is used as fuel enters the anode side, while an oxygen gas stream is used as oxidant at the cathode side. These cells are connected in series to give the demand voltage and the current.

3.2 Hydrogen Fuel Cells: Working Principles

The basic physical structure of a fuel cell (FC) is depicted in Fig. 3.1; the mode of operation depends on the presence of a membrane separator, and the catalyst, which is usually made of platinum powder and the carbon paper or cloth, is coated with a very thin layer of the platinum powder [1–3]. The platinum coats the sides of the plate, and it is placed next to the proton exchange membrane. Once hydrogen (fuel) is supplied to the cell, the platinum (catalyst) works on separating H_2 (which is the simplest atom consisting of just one proton and electrons) and is spilt into positive proton ion and negative charged electron. This operation occurs at anode side and known as oxidation of fuel. If the oxygen and hydrogen is mixed directly, the products will be water and heat due to this reaction being exothermic. This reaction needs to be controlled where the electrical current used to provide the load by damned electrical energy is created due to hydrogen oxidation resulting in two protons and two electrons.

The membrane separator allows protons to pass but not electrons. Therefore, electrons should pass only through the current collectors, and this produces direct electrical current ($I_{d.\ c}$). On the opposite side of the membrane (i.e. at the cathode

A. Albarbar, M. Alrweq, *Proton Exchange Membrane Fuel Cells*,
https://doi.org/10.1007/978-3-319-70727-3_3

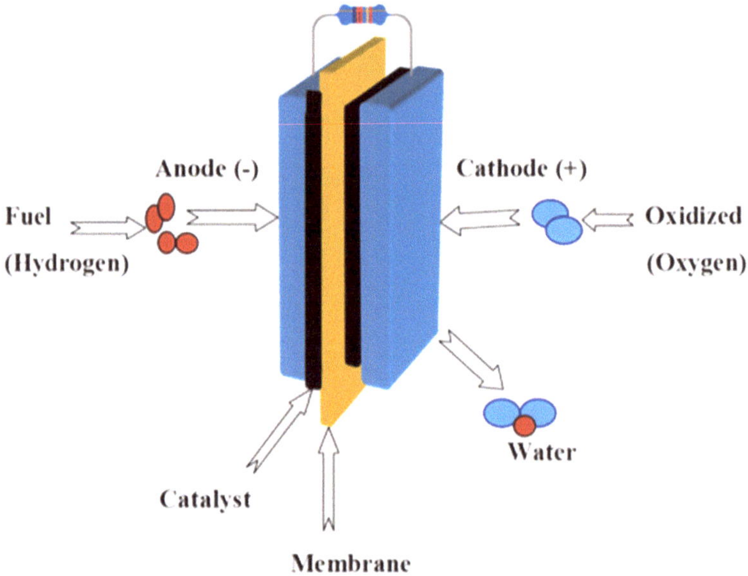

Fig. 3.1 The functionality of the proton exchange membrane fuel cell

side), the electron binds with the proton in the presence of a catalyst again, and with the presence of oxygen, the water, H_2O, is produced, and the heat spreads.

3.2.1 Proton Exchange Membrane Fuel Cells

Niedrach and Grubb invented the PEM fuel cells in 1960 at the General Electric Company. PEM fuel cell is the most commonly used type of FC; it is used in various fields due to its small size and its low temperature operation. PEM fuel cell uses a polymer as membrane (Nafion), which plays the role of mediator in the electrochemical cell, in a solid state, and thus it reduces the reaction temperature and increases the efficiency (speed of interaction in the start-up and responding when loading). The polymer separation membrane is positioned between two electrodes of perforated platinum, and it causes no risk of pollution due to its solid nature. Interaction takes place under 80°C, and when the membrane is exposed to water, it becomes conductive material for ions [4, 5]. The electrodes are made of platinum. The efficiency of such cells is up to 45–50%, and their power density is high compared with other types of cells where up to 350–600 mW/cm^2, and the reactions are defined in Eq. (3.1):

$$\text{at Anode}: 2H_2 \rightarrow 2H^+ + 4e^- \tag{3.1a}$$

$$\text{at Cathode}: O_2 + 4H^+ + 4e^- \rightarrow 2H_2O \quad (3.1b)$$
$$\text{Overall reaction}: 2H_2 + O_2 \rightarrow 2H_2O + \text{electricity} + \text{heat} \quad (3.1c)$$

Another type of polymeric cell runs directly on methane gas, and it differs from the first type in terms of the electrode's materials. These cells have a device for the preparation of fuel, where it works on the saturation of methane with hydrogen. The problem in this type of cell is the crossing of methane from the membrane, and there is currently a lot of research to address this issue [6].

3.2.1.1 Thermodynamic Analysis

Thermodynamics is the science that deals with the transformation of energy from one state to another. As the chemical energy is transferred to electrical energy in fuel cells, predictions that can be obtained using the equations of thermodynamics are essential to understand and model the performance of the fuel cell.

3.2.1.2 Proton Exchange Membrane Fuel Cell

The thermodynamic fuel cell is the study of the mechanism of electrochemical, thermal and electrical reactions within it, to indicate the performance and efficiency of the FC. Studying the topics mentioned based on the study of change in free energy of Gibbs, which is defined as the actual capacity of the production operation outside of the FC, without addressing the consumption of this energy to change the pressure and volume. It is the energy of the electrons in the external circuit, measured in KJ/mol and written as follows:

$$\Delta G_f = \Delta G_{f\,(\text{product})} - \Delta G_{f\,(\text{reactant})} \quad (3.2)$$

where the free energy results from the chemical reaction and the total energy enthalpy Δg_f minus the entropy losses Δ, and if we wrote Eq. (3.3) for one molecule:

$$\Delta g_f = \Delta g_{f\,(\text{product})} - \Delta g_{f\,(\text{reactant})} \quad (3.3)$$

Hydrogen and oxygen gases interact in FCs, and this reaction produces water so the equation becomes:

$$\Delta g_f = \Delta g_f(H_2O) - \Delta g_f(H_2) - \Delta g_f(HO_2) \quad (3.4)$$

The free energy that enters the cell is not totally converted into electrical energy. In other words, the total freed chemical energy in the electrochemical interaction, and that is equal to $f\Delta h$ (even if we consider that losses are neglected, such as energy used to change the pressure of gases (fuel)), is not equal to the amount of change in free energy of Gibbs transformed into electrical energy. This is due to, in a FC from each molecule of hydrogen on the two electrons 2e (pure hydrogen fuel), i.e. if there is a molecule of hydrogen and assume that the atoms of this molecule all interact when the negative pole (anode), the result which is $2e*N$ electrons, where N is the number of Avogadro. The load which connected with the external circuit gets $2e*N$ of electrons where e is the electron charge. The Faraday's constant, which is equal to (96,487 J/V-mole), is the result of the production of e in N of electrons:

$$2e*N = 2F \tag{3.5}$$

The variation of voltage between the cathode and the anode (potential energy) is E; thus we will have: Voltage x electric current =eclectic – $2FE$ [Joules].

This assumes that the reaction in the cell is reversible; therefore, as has been explained previously, the electrical supply is different in the free energy (Gibbs) resulting from the interactions that took place in the cell [7, 8]:

$$\Delta g^{-} = -2FE \rightarrow E = \frac{-\Delta g}{2F} \tag{3.6}$$

$$E_o = \frac{219100}{2*96487} = 1.135 \text{ volt} \tag{3.7}$$

In addition, this equation shows the open-circuit voltage of the FC and gives free energy in the data manual to a FC. Table 3.1 shows the values of the free energy, outcome and voltage of the open circuit in the event of the output water or water vapour to the cells and its fuel being hydrogen and oxygen. Figure 3.2 shows the entropy of hydrogen, oxygen and water as functions of temperature.

The voltage of the open circuit is associated with the real operating conditions at a temperature of operation, where the Gibbs energy changes when the fuel enters the cell, and operating temperature and water resulting from the interaction, based on the electrochemical reactions that take place in fuel cells [9].

The Nernst equation shows the relationship between the open-circuit voltage (E_o) in typical conditions ($T = 25^{\circ}\mathrm{C}$ $P_{H_2} = P_{O_2} = 1\,\mathrm{atm}$) and the operating voltage for FC (E) at different temperatures and pressures:

$$\Delta G = \Delta G_O + RT\mathrm{In}Q \tag{3.8}$$

Note that the FC with low temperature (PEM fuel cell and PAFC) needs to contain expensive metals to function as a catalyst to accelerate the reaction in the electrodes. Hydrogen, in general, is the only fuel used in these types of cells, whereas cells with high temperatures (MCFC and SOFC), the catalysts and fuel used are varied and diverse, so starting with the energy of a chemical reaction (Q) of the FC as in the equations:

Table 3.1 Free energy, yield stress and the open circuit if the output is water or water vapour

$H_2 + \frac{1}{2}O_2 \rightarrow H_2O$ 1 bar reactant pressure (standard condition)						
Temp (°C)	Product H_2O phase	$\Delta H_f\left(\frac{KJ}{mol}\right)$	$\Delta G_f\left(\frac{KJ}{mol}\right)$	$\eta\%$	$E_{O.\ C}(V)$	$E_{rev}(V)$
25	LHV(vapour)	−241.8	−228.6	94.5	1.253	1.185
	HHV(liquid)	−286	−237.3	83	1.482	1.229
80	LHV	−242.3	−226.2	93.3	1.256	1.172
	HHV	−283.8	−233.7	82.3	1.471	1.212
130	LHV	−242.8	−223.9	92.2	1.258	1.160
	HHV	−282.1	−230.4	81.7	1.462	1.195
200	LHV	−243.8	−219.1	89.8	1.259	1.131
600	LHV	−247.2	−198.1	80.1	1.277	1.023
1000	LHV	−249.4	−175.8	70.5	1.288	0.908

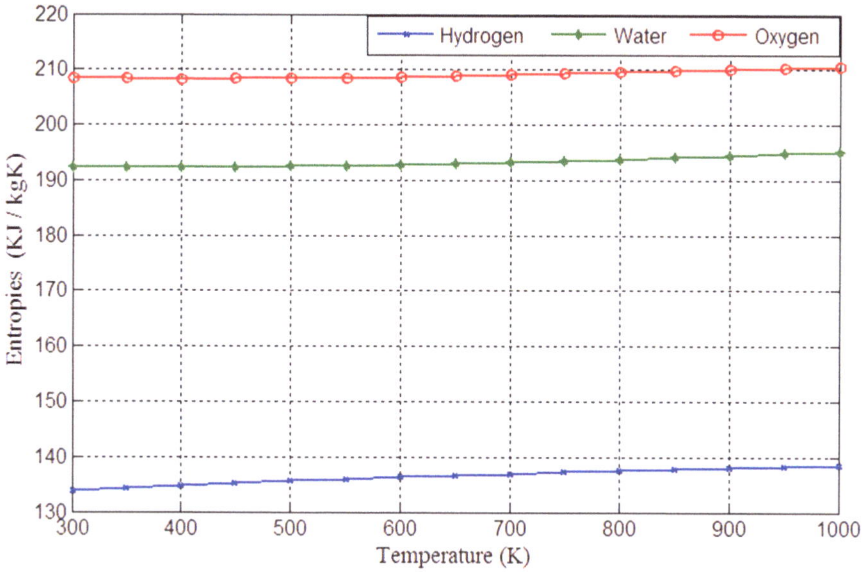

Fig. 3.2 The entropy of hydrogen, oxygen and water

$$aA + bB \rightarrow cC + dD \Rightarrow \text{where } Q = \frac{[C]^c[D]^d}{[A]^{ca}[B]^b} \tag{3.9}$$

$$\frac{1}{2}O_2 + H_2 \rightarrow H_2O \Rightarrow Q = [H_2]\frac{[O_2]^{\frac{1}{2}}}{[H_2O]} \tag{3.10}$$

$$E = E_o + \frac{RT}{2F}\left[\frac{\alpha\beta^{\frac{1}{2}}}{\delta}\right] + \frac{RT}{2F}\ln(p) \tag{3.11}$$

The above equations can be expressed as follow:

$$E = 1.23 - 0.9{*}10^{-3}(T - 298) + \frac{RT}{4F}\ln\left({P^2}_{H_2}{*}P_{O_2}\right) \tag{3.12}$$

When considering the pressure of the gases entering the cell and the water pressure generated by the electrochemical interaction and by dividing both sides of the equation by $-1/2F$ to become the Nernst equation depending on the equations mentioned earlier as follows:

$$E = E_o + \frac{RT}{2F} + \ln\frac{P_{H_2}{*}(P_{O_2})^{\frac{1}{2}}}{P_{H_2O}} \tag{3.13}$$

Equation (3.13) expresses the total pressure in terms of the partial pressure of gases, multiplied by a gas constant of each gas separately:

$$P_{O_2} = \beta p \;\; \& \;\; P_{H2} = \alpha P \& P_{H_2O} = \delta P \tag{3.14}$$

By substituting the value of P_{O_2} in Eq. (3.14), it will produce the final form of Eqs. (3.15) and (3.16) that expresses the influence of heat and pressure on the work of the cell as follows:

$$E = E_o + \frac{RT}{2F}\left[\frac{\alpha\beta^{\frac{1}{2}}}{\delta}\right] + \frac{RT}{2F}\ln(p) \tag{3.15}$$

$$E = 1.23 - 0.9{*}10^{-3}(T - 298) + \frac{RT}{4F}\ln\left({P^2}_{H_2}{*}P_{O2}\right) \tag{3.16}$$

Where:

R: is the gas constant (8.314 KJoules/ deg − mol).
T: is temperature in Kelvin, P the gas pressure in the anode and the cathode (bar).
F: is Faraday's constant (96,487 joules/volt-mole).
β: is the hydrogen gas constant and δ the water vapour constant.

3.3 Polarisation Phenomenon

When the FC is loaded, the voltage goes down by 60% or 70% of the open-circuit voltage. Both the output voltage and current density determine the characteristics of the *V-I* curve. The output voltage of PEM fuel cell is closely related to that thermodynamically predicted in FCs and three major losses, which are activation losses, ohmic losses and concentration losses. Activation losses occur due to electrochemical reaction. Ohmic losses occur due to ionic electronic conditions, while concentration losses are produced due to mass transport [10].

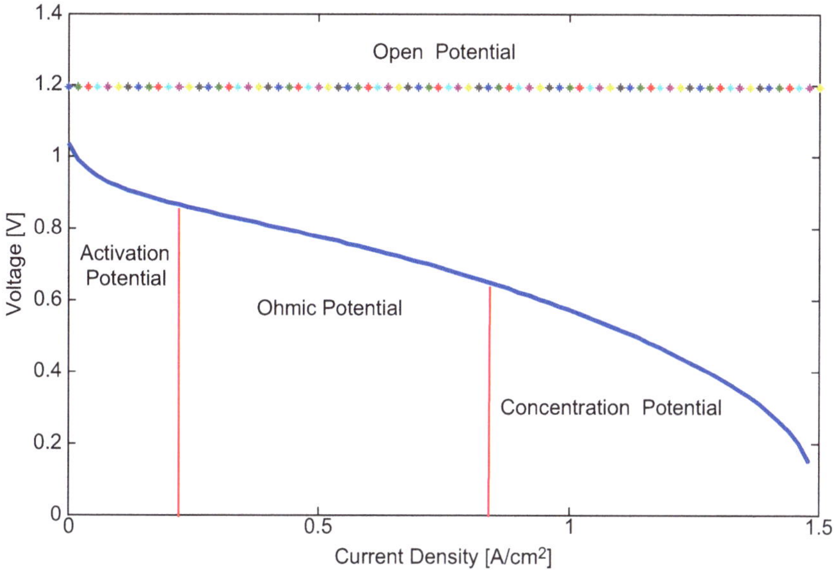

Fig. 3.3 The schematic diagram of the FC through the production cycle and three losses

It is obvious from Fig. 3.3 that the open-circuit voltage (E_O) is less than the theoretical value of the reversible voltage [1, 10] . This is proof that there is a loss in voltage even when there is no load connected to the FC.

The main reason for this loss is the fuel cross over and internal currents through the electrolyte through which only ions should transport. However, a small amount of fuel and less electrons will be conducted through the electrolyte. The electrolyte has a reducing effect on the open-circuit voltage, as depicted from the polarisation curve. When the current moves away from the zero point and the voltage initially and rapidly drops, this drop is due to the activation losses which are caused by the slowness of the reaction taking place on the surface of the electrodes. A small part of the voltage generated is lost since it drives the chemical reaction responsible for transferring the electrons to and from the electrode.

At medium current densities, the voltage losses become linear and decrease less slowly. The linearity characteristics represent the ohmic losses, which are also known as 'resistive losses', and they stem from the straightforward resistance to the flow of electrons in the various FC components, as well as the resistance to the flow of ions in the electrolyte. This voltage drop is linear, and it is proportional to the current density.

At higher current density, the voltage falls rapidly due to mass transport limitations in the cell. These are usually known as 'mass transport or concentration losses', and they result from the change in the concentration of the reactants at the surfaces of the electrodes.

The possible reasons for this change are:

1. The obstruction that prevents the gases from reaching the reaction sites on the membrane.
2. The accumulation of product water blocking the flow channels in the bipolar plates or the GDL. To avoid the drastic decrease in power density in this region, the optimal operating system for a FC is up to the maximum power density.

It is noteworthy that the terms used for losses are variable from one discipline to another; they are called losses, voltage drop or, conversely, overvoltage, which gives the impression that voltage increases instead of decreases, but it is the term usually used in electrochemical literature. In the following sections, each one of these losses will be described separately and presented in mathematical and graphical forms. The effects of varying certain operating conditions such as temperature and pressure on the performance of the fuel cell, as well as certain geometric dependant parameters such as the values of exchange current density and charge transfer coefficient, which are dependent on the electrode material and catalyst loading of the FC electrodes.

The analysis takes into consideration the following assumptions:

1. The FC operates under steady-state conditions.
2. The gases are assumed ideal compressible gases.
3. Due to the low velocity of the gases, their flow is laminar.
4. The product water is assumed to be in liquid form.
5. The electronic resistance of the FC components and the external circuit are constant over the range of operating temperatures.
6. The pressures of the anode and cathode are assumed to be the same.
7. The internal currents in the FC are equal to fuel crossover.

3.3.1 *Activation Polarisation*

When the cell is loaded, the nonlinear voltage drops suddenly by ($\eta_{act} > 50$–$100\,mV$) and then remains constant, because of the energy needed to trigger the interaction between fuel and oxidiser. The activation polarisation (η_{act}) is nonlinear, and it has large values at low current density. Both electrochemical and chemical reactions are very similar since they involve an activation barrier that needs to be overcome by the reacting species. Consequently, part, of the electrode potential is lost since it drives the electron transfer rate to the one required by the current demand (Table. 3.2).

The reaction kinetics is expressed by the Butler–Volmer equation where the mass transfer effects are neglected [11, 12]:

$$i = i_0\left[e^{\frac{\alpha n\eta_{actc}}{RT}} - e^{\frac{1(1-\alpha)nF\eta_{acta}}{RT}}\right] \tag{3.17}$$

Table 3.2 Values of constant parameters that utilised to plot the polarisation curve in Fig. 3.3

Parameter	Value	Units
E_o	1.23	Volt
b	0.032	kJ. mole/kg. °C
R	$3*10^{-3}$	Ohm
α	0.3	–
$P_{O_2} = P_{H_2}$	2	atm
T	353	Kelvin

Where:

R: the universal gas constant (8.314 kJ/kmol. K).
T: the temperature of cell operation in Kelvin.
n: the number of electrons involved per mole of electrolysed component.
α: the charge transfer coefficient, which is the portion of the electrical energy assisting the forward reaction, $1 - \alpha$: hinders the reverse process.
The α value depends on the reaction involved and the material the electrode is made from, and it must be in the range of ($0 < \alpha < 1$)).
F: Faraday's constant(96485 C/mole).
η_{actc} and $\eta_{acta}\eta$: the activation polarisations on the cathode and anode, respectively.
i: the current density (A/cm^2).
i_o: the exchange current density, which is the rate of flow of electrons to and from the electrolyte.

In Eq. (3.17), the first exponential term expresses the forward reaction voltage (the reduction reaction on the cathode), while the second term represents the backward reaction voltage (the oxidation reaction on the anode). The equation can be written in the logarithmic form as:

$$\ln\left(\frac{i}{i_o}\right) = \frac{-\alpha nF\eta_{actc}}{RT} + \frac{(1-\alpha)nF\eta_{acta}}{RT} \tag{3.18}$$

If the activation overvoltage on the cathode is considerably higher than that on the anode, the first part of the equation which represents the forward oxidation reaction prevails, and the second part can be ignored and vice a versa in the case of prevailing anodic voltage.

Tafel's law represents these two expressions separately for anode and cathode. To express the net cathodic overvoltage, the backward activation overvoltage (anodic reaction) is negligible, and the equation becomes:

$$\eta_{actc} = -\frac{RT}{\alpha nF}\ln\left(\frac{i_c}{i_o}\right) \tag{3.19}$$

The 'exchange current density' can be defined as 'the current flowing equally in each direction at the reversible potential'. The chemical reaction's 'equilibrium' is initiated when the forward and backward rates are the same.

For electrochemical reactions, the rates of reaction define the current which, at equilibrium, would be the exchange current density. For a net, anodic overvoltage, the anodic reaction becomes the forward reaction, and the cathodic reaction can be neglected; Eq. (3.20) becomes:

$$\eta_{\text{actc}} = \frac{RT}{(1-\alpha)nF} \ln\left(\frac{i_{\text{c}}}{i_{\text{o}}}\right) \tag{3.20}$$

Subscripts a and c utilised in Eqs. (3.19) and (3.20) are used to express the anodic and cathodic current densities, respectively. Both equations are in the form known as Tafel's law, which can be written in terms of current density as follows:

$$\eta_{\text{actc}} = x + y \ln i_{\text{c}} \tag{3.21}$$

This equation can also be written in terms of both current density and exchange current in the following form:

$$\eta_{\text{act}} = b\log\left(\frac{l_{\text{c}}}{i_o}\right) \tag{3.22}$$

where b in this equation is equal to 'y' as follows:

$$b = \frac{RT}{\alpha nF} \tag{3.23}$$

Tafel's equation can be used in situations where the activities of the species involved in the reaction are not very much affected by the current flow. This equation can be utilised to deduce the exchange current density which takes place when $\eta_{\text{act}} = 0$.

Equation (3.22) can be rearranged to give '$\ln i_{\text{c}}$' in terms of 'η_{act}' and considering the other terms as constant gives:

$$\eta_{\text{act}} = -\frac{RT}{\alpha nF} \ln i_{\text{o}} + \frac{RT}{\alpha nF} \ln i_{\text{o}} \tag{3.24}$$

Writing the equation using the coefficient 'b':

$$\eta_{\text{actc}} = b \ln (i_{\text{o}}) - b \ln (i) \tag{3.25}$$

If proper values were substituted in Eq. (3.25) and putting $n = 4$ for the cathode side and $\alpha = 0.5$ [106] and the operating temperature $T = 373$ K, the corresponding value of 'b' $= 0.0161$. Figure 3.4 depicts the equation plotting for various exchange current densities.

It is obvious from the Fig. 3.4 that increasing the exchange current density results in reducing the activation losses. A typical value for the exchange current density (i_{o}) would be about 0.1 mA/cm^2 at the cathode and about 200 mA/cm^2 at the anode (for FC running on air at ambient pressure and low temperature). This

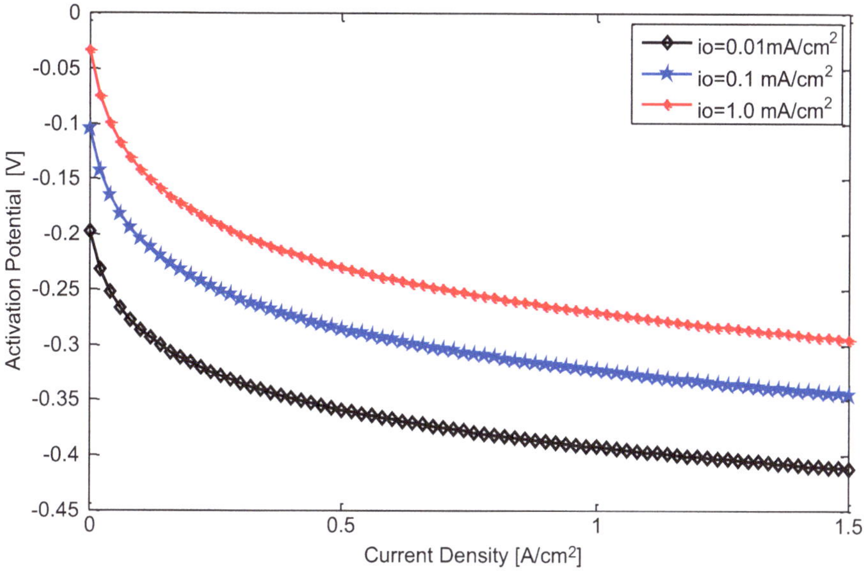

Fig. 3.4 The impact of exchange current density on the activation potential

makes losses due to activation much greater on the cathode side, and therefore it requires more catalyst loading to improve the kinetics of the reaction.

It can also be seen from the Fig. 3.4 that the current density has an adverse effect on the voltage, as it drops exponentially as the current density increases. Jaouen et al. conducted experiments on a PEM fuel cell at a temperature of 50°C [13]. The study found that there is a linear relationship between the logarithm of the exchange current density i_0 and the logarithm of the oxygen partial pressure, according to:

$$i_0 = x.e^{(yP_{O2})} \tag{3.26}$$

where $x = 1.27 \times 10^{-8}$ and $y = 2.06$. However, this relationship is only applicable to the conditions of the experiment performed in the study, and hence it is not applicable to other fuel cells. The exchange potential particularly at the cathode side is a mixed potential due to reactions. In addition, there are other geometric variables and operating conditions that contribute to the value of the exchange current density. The study [14] also concluded from the experiments that the oxygen partial pressure is a factor in determining the value of 'i_0'. To understand the impact of the charge transfer coefficient on the activation overvoltage, Tafel's equation is plotted for different values of the charge transfer coefficient (α) for a given value of exchange current density $i_0 = 0.01$ and an operating temperature of $T = 373$K. (Fig. 3.5)

As the charge transfer coefficient decreases, the activation overvoltage increases exponentially. This shows the importance of the charge transfer coefficient which depends on the type of electrode material; therefore, the type of electrode material

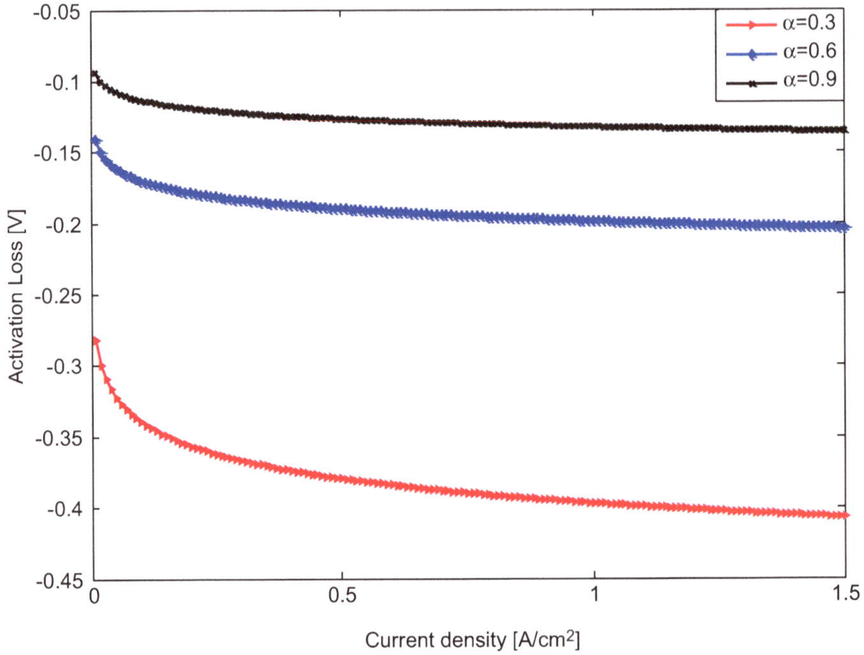

Fig. 3.5 The effect of varying the values of α in the activation overvoltage, for exchange current density ($i_o = 0.01$) and operating temperature $T = 373$ K

is an important factor in improving the power output of the fuel cell. When the same equation is plotted with various operating of both temperatures and current densities, it is obvious that the activation losses increase at higher temperatures. However, this effect is balanced by the increased activities of the reactants due to higher temperatures, in accordance with the kinetic theory of gases. For PEM fuel cell, with effective catalyst action, the hydrogen electrode operates close to thermodynamic equilibrium conditions. The hydrogen oxidation reaction proceeds readily so that the anodic activation potential; 'η_{acta}' is negligible.

In contrast, at the thermodynamic equilibrium potential, the rate at which the oxygen is reduced, in aqueous media, is around (5–10) times slower than that of hydrogen, even with the best catalysts currently available. Thus, the cathodic reaction activation can be attributed to the voltage drop. The second exponential term in Eq. (3.24) that expresses the anodic part of the total activation potential can be ignored as reported in most studies made of MEMFCs. Therefore, the equation can be simplified to Tafel's law.

However, this is not applicable for other types of FCs, such as the direct methanol FC (DMFC) which is like PEM fuel cell in using the same type of membrane electrolyte with a different catalyst and the same construction, but the activation over voltages on the anode are significant and should be taken into consideration in the equation.

Many methods were proposed to reduce the activation polarisation, such as [1, 13, 14]:

1. Increasing cell temperature
2. Doping the electrode with effective catalysts
3. Increasing the surface area of the electrode by making it rougher
4. Increasing the reactant concentration of air
5. Increasing the pressure (through increasing catalyst site occupancy which increases the open-circuit voltage)

3.3.2 Ohmic Polarisation

All materials have internal (or intrinsic) resistance to the flow of charges. This resistance causes ohmic polarization, which leads to the loss of cell voltage. The components of the FC, including the electrolyte, the catalyst layer, the gas diffusion layer (GDL), bipolar plates, interface contacts and terminal connections, contribute to its electrical resistance. The resulting voltage reduction is known as the 'ohmic loss', and it includes the electronic (R_e) and ionic (R_i) contributions to FC resistance. This can be written as Eq. (3.27):

$$\eta_{ohm} = IR_{internal} = I(R_e + R_i) \tag{3.27}$$

Since the transportation of the protonic is more difficult than electronic charge transportation, the R_i dominates the reaction. In Eq. (3.27) R_i represents the ionic resistance of the electrolyte, while the R_e represents the total electrical resistance of all conductive FC components, which includes bipolar plates, cell interconnects and contacts.

The ohmic polarisation (η_{ohm}) is positively proportional to the high-current density and internal resistance such as the electrode resistance, the bulk electrolyte (membrane) resistance and interface contact resistance between electrodes and electrolyte. Since the cell's resistance is almost constant, the ohmic polarisation changes linearly, and this is due to the emergence of resistance while crossing of ions in the electrolyte and ohmic resistance electrodes.

The reduction of these resistances by using the appropriate electrolyte and metals in the electrodes can be adopted to overcome this problem. In addition, the ohmic losses may be reduced by using high-conductive membranes as well as reducing the space between electrodes.

The ability of a material to support the charge flow through it is known as conductivity.

The electrical resistance of FC components is often referred to as conductance (s) in the state of the area, which is the inverse of resistance:

$$\sigma = \frac{I}{R_{internal}} \tag{3.28}$$

Resistance characterises the size, shape and properties of the material, as expressed by Eq. (3.29):

$$R = \frac{\rho_{\text{ele}} l}{A} \tag{3.29}$$

Where:

$$\rho_{\text{ele}} = \frac{1}{\sigma}$$

Then Eq. (3.29) can be rewritten as follows:

$$R = \frac{l}{\sigma A} \tag{3.30}$$

where ρ_{ele} is the electrical conductivity (ohm^{-1} cm^{-1}), l is the length (cm) of the conductor and A is the cross-sectional area (cm^2) of the conductor.

The current density, which expresses the number of the electrons in a square metre, can be expressed by the following equation:

$$i = \frac{I}{A} \tag{3.31}$$

Current density (I) can also be determined by the number and the average velocity of carriers as in Eq. (3.32):

$$i = Qn_{\text{carriers}} v = \sigma \xi \tag{3.32}$$

where A is the active area of FC, n_{carriers} is the number of charge carriers (carriers/cm^3), Q is the charge of the carrier (1.6×10^{-19}), v is the average drift velocity (cm/s) with which the charge carriers move and ξ is the electric field.

The final equation for conductivity is:

$$\sigma = Qn_{\text{carriers}} \frac{v}{\xi} \tag{3.33}$$

The last part of Eq. (3.33) is known as the mobility. Two main factors can characterise an equation for material conductivity in a specific way; these are:

1-. The number of carriers available
2-. The mobility of those carriers in the material, which can be written as:

$$\sigma_i = (|Z_i| * F) * C_i * u_i \tag{3.34}$$

where C_i is the number of moles of charge carriers per unit volume, u_i is the mobility of the charge carriers within the material, Z_i is the charge number (valence electrons) for the carrier and F is Faraday's constant.

The reduction of FC resistance will lead to FC performance improvement, since the FC resistance changes with area. When studying ohmic losses, it is essential to compare resistances on a per-area basis using the current density. Ohmic losses can be determined from current density using Eq. (3.35):

$$\eta_{\text{ohm}} = i(\text{AS}\, R_{\text{ohmic}}) = i(A* R_{\text{ohmic}}) \tag{3.35}$$

where AS R_{ohmic} is the area-specific resistance of the PEMF. The conduction mechanisms are different for electronic versus ionic conduction. In a metallic conductor, valence electrons associated with the atoms of the metal become free to move around in the metal.

In a typical ionic conductor, the ions move from site to site, hopping to ionic charge sites in the material. The number of charge carriers in an electronic conductor is much higher than an ionic conductor. Electronic and ionic transportation are shown in Figs. 3.6 and 3.7, respectively.

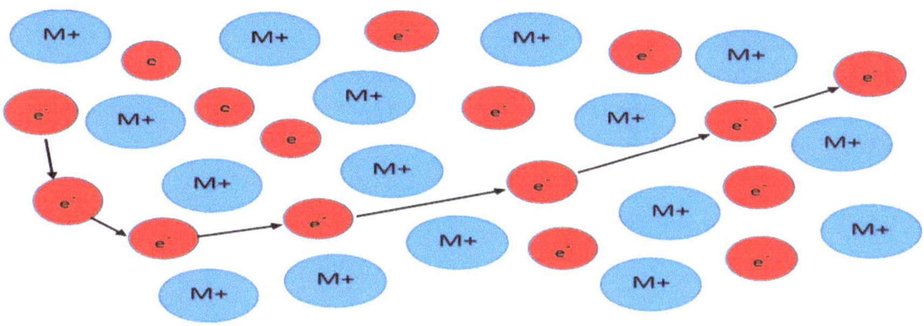

Fig. 3.6 The transportation of the electrons in a metal

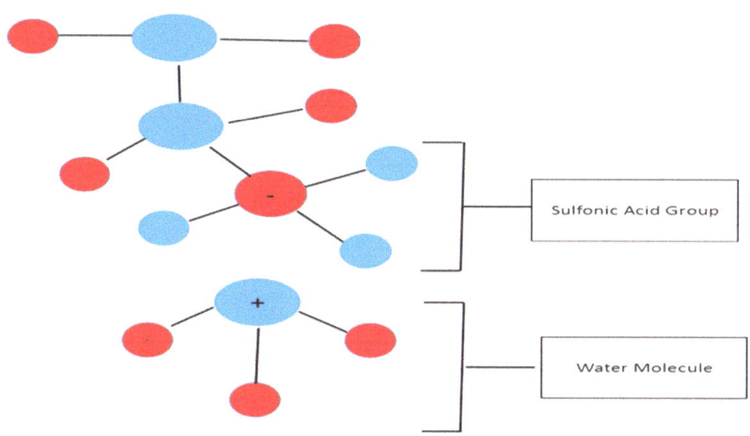

Fig. 3.7 Ionic transportation in an electrolyte

The losses from ohmic contact can be reduced by either increasing the active area or decreasing the channels area. This will help to optimise the voltage and current and ultimately improve the performance and efficiency of the FC. The most effective way to reduce ohmic loss is to use either a better ionic conductor for the membrane layer or a thinner membrane layer.

Thinner membranes are advantageous for PEM fuel cells because they guarantee the saturation of the anode electrode through 'back' diffusion of water from the cathode. For very high-current densities (which means fast fluid flows), mass transportation causes a rapid drop off in the voltage. Oxygen and hydrogen cannot diffuse through the electrode and ionise quickly enough; therefore, products cannot move out at the necessary speed. Ionic resistance in the electrolyte mainly causes ohmic over potential for the FC, and this can be expressed as:

$$\eta_{\text{ohm}} = iR_{\text{internal}} = iA\left(\frac{\delta_{\text{thick}}}{\sigma A}\right) = \frac{\delta_{\text{thick}}}{\sigma} \tag{3.36}$$

where A is the active area of FC, δ_{thick} thick is the thickness of the electrolyte layer, and σ is the conductivity.

It is obvious from Eq. (3.36) and Fig. 3.8 that the ohmic potential can be reduced by employing a thinner electrolyte layer or a higher ionic conductivity electrolyte (Figs. 3.9 and 3.10).

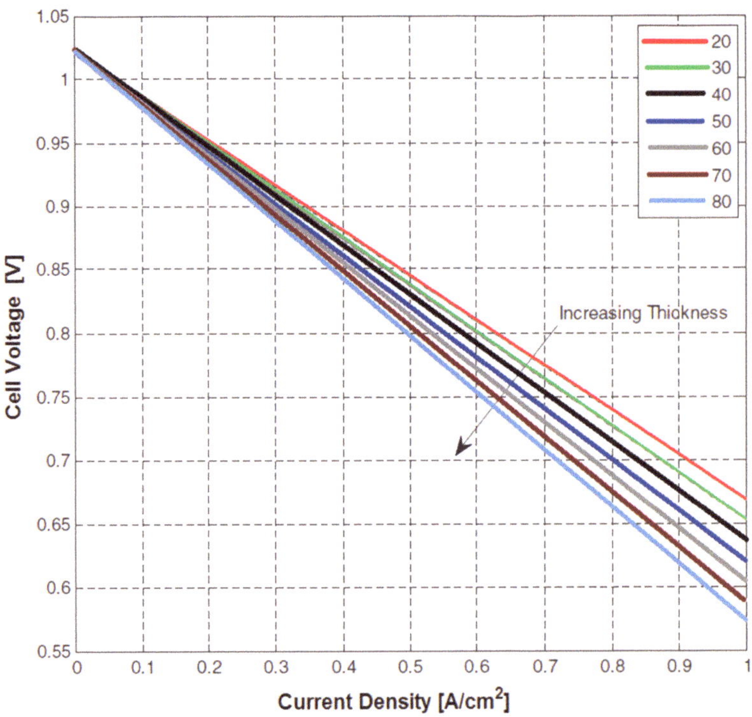

Fig. 3.8 Fuel cell voltage and current as a function of the channel length

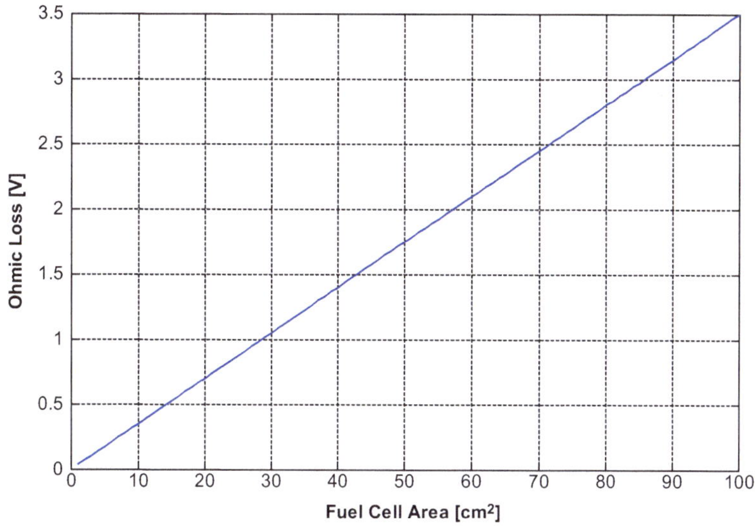

Fig. 3.9 Ohmic losses as a function of fuel cell area

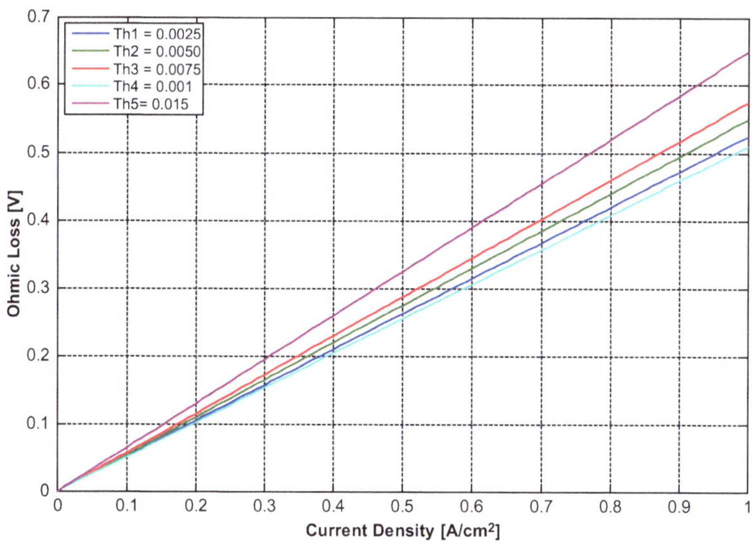

Fig. 3.10 Ohmic losses as a function of membrane thickness

Figure 3.11 shows the local conductivity as a function of membrane thickness. The total ohmic losses for the FC can be written as:

$$\eta_{\mathrm{ohm}} = iA\sum R = IA\left[\frac{I_a}{\sigma_A A} + \frac{I_e}{\sigma_e A} + \frac{I_c}{\sigma_c A}\right] \tag{3.37}$$

where I is the length or thickness of the material.

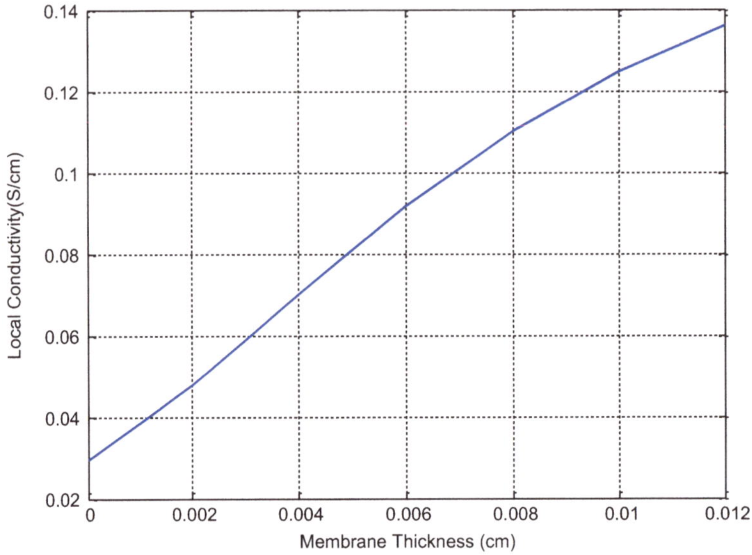

Fig. 3.11 Membrane thickness and local conductivity

The first term in Eq. (3.37) applies to the anode, the second to the electrolyte and the third to the cathode. In the bipolar plates, the 'land area' can vary depending on the flow channel area. When the land area is reduced, the contact resistance increases accordingly, since the land area is the term in the denominator of the contact resistance.

3.3.3 Concentration Polarisation

The concentration polarisation, which also is known as 'mass transport losses', occurs when there is a change in the concentration of one of the reactants at the surface of the electrodes, and this occurs when a chemical species contributes to the reaction (in a small amount) due to obstruction in the pathway of this species. This loss type is also known as the concentration polarisation. This phenomenon is also known as 'Nernstian' as it is related to the concentration effects modelled by Nernst equation.

The reduction of reactants concentration reactants on the rate at which they are being consumed. This concentration reduction depends on the current drawn from the FC as well as on the physical characteristics of the system.

Obviously, these factors are dependent on the variation in pressures, concentrations of the gases and to the gas transportation rate (the rate at which gases move from the flow channel to the surface of the membrane through the catalyst layer and the gas diffusion layer (GDL)).

The initial concentrations of the reactant gases, represented by their partial pressures, have an effect on open-circuit voltage. The pressure of the gases decreases in the FC while being consumed, until the point where the amount of reactant gases reaching the electrolyte is equal to the rate of their consumption. At this point, the partial pressure of the reactant gas is equal to zero. Therefore, it is not possible to increase the current output of the FC beyond this point, as it has reached its maximum current density. This density is termed as 'limiting current density'. Since the pressure is assumed to have dropped to zero at the limiting current density linearly due to mass transport, the initial pressure at zero current was assumed to be. It can be seen from the similarity of triangles depicted in Fig. 3.11 that by substituting this value in the Nernst Eq. (3.16), which expresses the relationship between the voltage drop and partial pressures of the reactant gases, the following expression is obtained [13] (Fig. 3.12):

$$\eta_{\text{conc}} = -\frac{RT}{nF}\left\{1 - \frac{i}{i_1}\right\} \tag{3.38}$$

where:

n: is the number of electrons transferred per molecule in the reaction, in the case of hydrogen–oxygen FC $n = 2$ for hydrogen, and $n = 4$ for oxygen.
R: is the universal gas constant (8.314 KJ/kmol).
T: is the temperature of operation in Kelvin.
F: is Faraday's constant.

This can be compared to Eq. (3.23) where:

$$b = -\frac{RT}{\alpha nF} \tag{3.39}$$

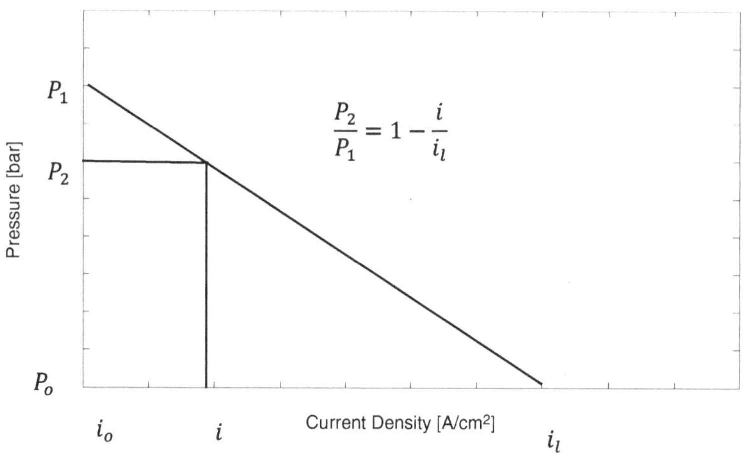

Fig. 3.12 Assumed variation of current density with concentration pressure

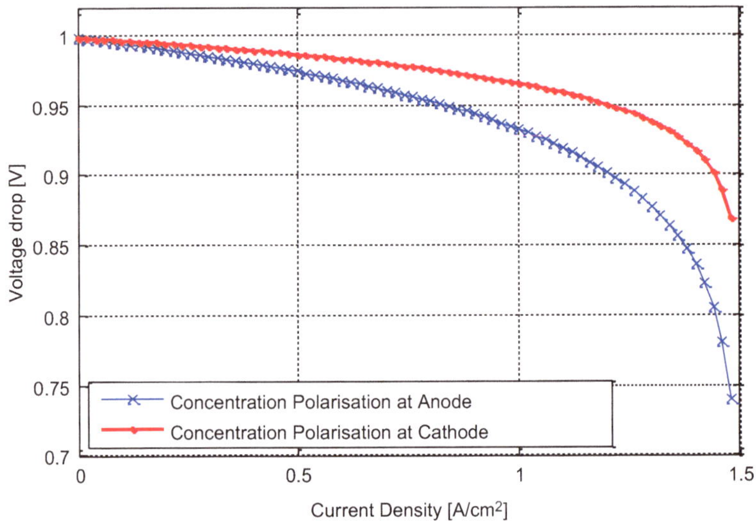

Fig. 3.13 Concentration overvoltage at the anode and cathode at 353 K, open-circuit voltage is taken as 1 volt

Therefore, Eq. (3.38) can be written as:

$$\eta_{\text{conc}} = \alpha \ln \left\{ 1 - \frac{i}{i_1} \right\} \tag{3.40}$$

Equation (3.39) can be employed to characterise the concentration potential for H_2 and O_2 using appropriate values of 'n' at a temperature of 353 K and charge transfer coefficient '$\alpha = 0.5$' [14]; the corresponding plots are depicted in Fig. 3.13, where the open-circuit voltage is taken as 1 Volt.

In Fig. 3.13, the effect of the concentration polarisation is significant at the anode compared to the cathode. This is due to the significant speed of the reaction kinetics on the anode. Meanwhile, any effect that results in a delay to the reaction will be more noticeable on the anode. For 'limiting current', it does not occur immediately, and the curve decreases gradually at the anode side. Whereas, at the cathode side, the curve's drop-down is more gradual. Both curves drop down to the limiting current value simultaneously.

Therefore, the final drop at the cathode side is more rapid, and it is not possible to draw more current from the FC beyond this point since this attempt will result in a sharp drop in cell voltage, which decreases the power output.

Kim et al. [15] presented different methods to model the concentration potential through the empirical equation based on experimental data, and this equation is expressed as follows:

$$\eta_{\text{conc}} = m\exp(ni) \tag{3.41}$$

The parameters 'm' and 'n' were not physically interpreted; however, the interpretation was that 'm' correlates to the electrolyte conductivity and 'n' to the porosity of the gas diffusion layer.

Consequently, it can be concluded that both 'm' and 'n' relate to water management issues. The partially dehydrated electrolyte membrane causes the decrease in conductivity, and this can be represented by 'm', while the extra liquid water leads to a reduction in porosity and hence to an early onset of mass transport limitations, which can be captured by the parameter 'n' [16, 17].

It is noteworthy that the equation is derived based on curve fitting techniques, and thus it can only be applied to the FC and operating parameters of the particular experiments carried out by the authors [18].

Typical values for 'm' and 'n' parameters are $= 2.11e^{-5}$ and $n = 0.008$, as recommended in [19].

Another parameter that can be taken into consideration to minimise the concentration losses is the FC's geometry. This parameter helps in increasing the value of the limiting current and improving the range of operation of the FC, that is achieved through reducing pressure drop in the flow channels and increasing the active area of the membrane and improved water management which removes the water produced by the reaction at the cathode side, to maintain access for the reactants to reach the active sites on the membrane.

3.3.4 Overall Voltage

The above three kinds of overvoltage which are activation, ohmic and concentration incorporate throughout the operation modes of FCs. The effect of that overvoltage will cause the nonlinearity of the performance curve of the FC which is also known as the polarisation curve, plotted with respect to the current required from FC.

To visualise the combined effect of these losses, a general equation that expresses the summation of their basic equation captures the total impact of the losses. The basic equation is:

$$E = E_o - \eta_{act} - \eta_{ohm} - \eta_{con} \tag{3.42}$$

3.3.5 Notes and Observations on Polarisation Curve

The following notes can be concluded from the graph:

1. Figure 3.3 presents the typical shape of the FC's performance curve. However, under practical conditions, some deviation from this shape is expected, due to the idealistic assumptions made during derivation of the equation and does not consider all the factors affecting the practical applications.

2. This study applies only to the changes in both pressure and temperature and their effect on the performance of the FC. However, these changes will also affect the conductivity of the various components, viscosity of the fluids. In addition, some parameters that were assumed constant, such as the specific heat at constant pressure, will also vary.

3.4 Hydrogen Fuel Cell Efficiency

3.4.1 Current Efficiency

The current efficiency (η_i) can be defined by the following equation:

$$\eta_i = I_{\text{cell}}/n\,N{*}F \quad (3.43)$$

Where:

I_{cell}: actual cell current
n: electrons involved in the reaction
F: Faraday's constant
N: molar rate of reaction of the fuel (mol/sec)

The current efficiency can be expressed as a percentage of current flowing through an electrolytic cell (and electrode), to complete the desired chemical reaction, compared with the ideal case. For example, in the FC that used hydrogen fuel, ideally every molecule of hydrogen (H_2) at the interaction will produce two electrons, which contributes to the current flow. The inefficiencies arise from reactions other than the intended one taking place at the electrodes or the side reactions consuming the current. For instance, some hydrogen molecules may go through the electrolyte with react at all, or the hydrogen molecules do react but the current produced is driven through the electrolyte and never contributes to the current flow.

3.4.2 Thermal Efficiency

The thermal efficiency of the fuel cell is known as:

$$\eta_{\text{thermal}} = \frac{P_{\text{out}}}{P_{\text{input}}} \quad (3.44)$$

Where:

ζ_{thermal}: the thermal efficiency
P_{Out}: power output
P_{input}: power input

Efficiency is a term that may refer to the process efficiency or the system efficiency. Process efficiency indicates how efficient a process in the system is performing. This definition does not express the total system efficiency, such as the combustion process in a heat engine which could reach 95%, while the system efficiency is in the range of 28%. The efficiency of the thermodynamic process occurring in the fuel cell is the theoretical maximum efficiency allowed by the second law of thermodynamics and can be written as:

$$\eta_{\text{thermal}} = \frac{\Delta G_T}{\Delta H^o} \tag{3.45}$$

where ΔG_T is the Gibbs free energy at the cell operating at temperature T is ΔH^o is the reaction enthalpy at the standard temperature and pressure 101.3 Pa, 298 K.

For PEM fuel cells (and other FCs) that utilise hydrogen as fuel, it is vital to determine the proper value of enthalpy depending on the phase at which water is produced.

Higher heating value HHV is used when the water product is liquid at 298 K (25° C), and lower heating value LHV is used when the water product is vapour at 423 K (150°C), due to the use of part of the enthalpy in converting water to vapour. It is noteworthy that in PEM fuel cell, the water product is produced at 353 K (80°C), and the calorific value of the fuel should be between the lower and higher heating value, as indicated in the definitions of the heating values.

From standard thermodynamic values, the values for the Gibbs free energy 'ΔG' for hydrogen and the enthalpy of the reaction 'ΔH' at 353 K (80°C) are approximated by linear interpolation, Table 3.3. Applying Eq. (3.44) for thermal efficiency above, using the interpolated values at 353 K (80°C) from Table 3.3, it follows that the maximum thermal efficiency is approximately equal to 0.87. Hence, from the equation of thermodynamic efficiency above, it can be concluded that:

$$\Delta G_{\text{T}} = 0.87 * \Delta H^o \tag{3.46}$$

Gibbs free energy expresses the maximum thermodynamic output possible in an electrochemical process. However, in real operation, the actual power output derived from the FC is:

The following equation gives actual electrical power output:

$$P_{\text{cell}} = VI \tag{3.47}$$

Table 3.3 Gibbs free energy, enthalpy and calorific value for hydrogen

Value	ΔG_{T} KJ/mole	ΔH^o KJ/mole	Calorific value MJ/Kg
LHV	223	240.4	120.21
HHV	237.1	285.8	142.18
Interpolated value at 353 K	230.5	264.6	132

To determine the FC's electrical efficiency, the actual output should be compared to the actual input, which is the total enthalpy of the reaction, hence:

$$\eta = \frac{\text{Actual electrical output power}}{\bar{m}\Delta H^o} = \frac{IV_{\text{cell}}}{\bar{m}\Delta H^o} \tag{3.48}$$

The fuel cell's current is equal to:

$$I = n{*}\bar{m}{*}F \tag{3.49}$$

Where:

$\bar{m}$: the molar flow rate of fuel
n: the number of electrons transferred per molecule in the reaction
V_{cell}: the measured cell voltage

Hence, from Eq. (3.47), the value of ΔH^o can be rewritten as below:

$$\mid \Delta H^o \mid = \frac{\mid \Delta G_T \mid}{0.87} = \frac{\mid -nFE_o \mid}{0.87} \tag{3.50}$$

Absolute numerical values are considered for efficiency calculation (since the negative sign indicates the direction of energy transfer). The substituting Eqs. (3.47) and (3.48) in (3.49) yields the electrical efficiency of the FC:

$$\zeta = 0.87\frac{V_{\text{cell}}}{E_0} \tag{3.51}$$

where V_{cell} is the measured cell voltage, which is a function of current density, and E_0 is the reversible voltage of the FC.

The expression can be interpreted as the maximum theoretical efficiency multiplied by the electrical efficiency, i.e. (the $\eta = \eta_{\text{thermal}}{*}\eta_{\text{ele}}$).

Following the same lines, the electrical efficiency is the ratio of measured electrical output to actual electrical input, which can be written as:

$$\eta_{\text{ele}} = \frac{iV_{\text{cell}}}{(i + i_{\text{int}}){*}E_o} \tag{3.52}$$

Where:

i: the current density
i_{int}: the cross over current which is assumed to be equivalent to internal currents (both currents are defining the input power together with the theoretical reversible voltage of the FC). From Eq. (3.50) and the definition of maximum thermal efficiency:

$$\eta = \frac{0.87{*}iV_{\text{cell}}}{(i + i_{\text{int}}){*}E_o} \tag{3.53}$$

This relationship is plotted in Fig. 3.13.

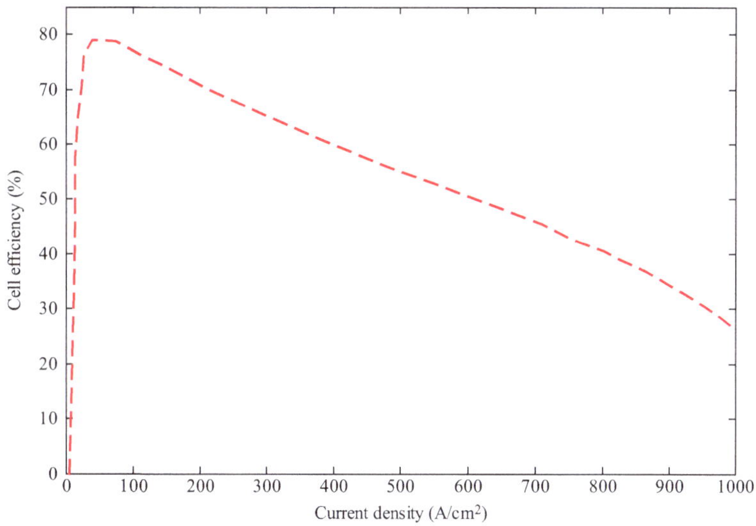

Fig. 3.14 Efficiency as functions of current density

It is clear from Fig. 3.14 that the efficiency of the FC is higher at low fuel flow rates, which corresponds to low current densities. The efficiency curve has similar behaviour to that of the voltage curve. More effort is needed to improve FC voltage and reducing voltage losses to enhance the performance of the FC.

According to the definition of efficiency in thermodynamics, another approach can be used to define the FC's efficiency, namely, by comparing the actual output with the input calorific value.

In this case, the system efficiency of the FC can be written as:

$$\eta_{FC}\% = \frac{V_{cell} * I}{mH_2 * \text{Calorific value(LHV)}} * 100 \tag{3.54}$$

Using the expression for current, for a hydrogen FC:

$$I = \frac{mH_2 * 2F}{MH_2} \tag{3.55}$$

Equation (3.54) can be rewritten after the compensation for the value of I as follows:

$$\eta_{FC}\% = \frac{V_{cell} * 2F}{MH_2 * \text{Calorific value(LHV)}} * 100 \tag{3.56}$$

By substituting the values for Faraday's constant, molar mass of hydrogen and the interpolated calorific value for hydrogen, the efficiency of the FC becomes:

$$\eta_{FC}\% = \frac{V_{cell}}{1.38} * 100 \tag{3.57}$$

Comparing the two expressions for efficiency as in Eqs. (3.54) and (3.55), the theoretical open-circuit voltage of a pure hydrogen FC can be found as follows:

$$\eta_{FC}\% = \frac{V_{cell}}{1.38} = 0.87\frac{V_{cell}}{E_o} \tag{3.58}$$

Therefore, the reversible voltage of the FC if the water product is steam at 80 °C is:

$$E_o = 0.87 * 1.38 = 1.2006 \text{ V} \tag{3.59}$$

This value is almost equal to the value of potential of equilibrium for hydrogen FCs which is 1.229 V, and this proves the above method for calculating the efficiency of the FC.

3.4.3 Fuel Utilisation Efficiency

There is a significant issue in FCs which is the venting of the fuel out of the cell. However, this issue can be overcome by operating the FC dead ended or in the case where the excess fuel is recalculated or used in another process. In this way the fuel is utilised properly.

The final efficiency of the FC has to be multiplied by the fuel utilisation value to calculate the exact efficiency of the FC. By assuming fuel utilisation efficiency was given the term 'ψ', the above expression of efficiency Eq. (3.56) becomes:

$$\eta = \psi * \frac{iV_{cell}}{(i + i_{int}) * E_o} \tag{3.60}$$

3.5 Summary

The main features and functionality of PEM fuel cells were presented starting with the contents and basic components of fuel cell and the fundamentals of PEM fuel cells. A thermodynamic investigation followed to establish the relationship between current and voltage and their relationship with other operations, such as pressure, temperature, exchange current density, charge transfer coefficient and gas concentrations in the FC. The analysis based on the polytrophic index is used to study the effects of pressure and temperature on FC performance, which shows the effects of irreversibility on output voltage of the FC.

A comprehensive study and analysis of FC take into consideration the actual operating conditions, internal currents, fuel utilisation efficiency and thermal and electrical efficiencies that involve expressions for the efficiency of the FC, and the formula for FC efficiency is derived and employed to plot the complete curve of efficiency against current density.

All derived equations can be utilised in performing parametric studies on FC performance.

The graphical representations of the solutions of the derived equations would help in finding the optimum combination of the design variables for changing operating conditions. They can also be employed in making a more complex numerical model of the system which can be resolved using computational methods to simulate the FC performance and then finding opportunities for optimisation.

The knowledge and findings acquired in this study will be useful in understanding the parameters that have an influence on PEM fuel cell and formulating the mathematical model, which will be used to study the performance and efficiency issues of fuel cell systems.

References

1. Alrweq, M., Albarbar, A. (2016). Investigation into the characteristics of proton exchange membrane fuel cell-based power system. IET Science, Measurement & Technology. doi: https://doi.org/10.1049/iet-smt.2015.0046 , Online ISSN 1751–8830.
2. Laberty-Robert, C., Valle, K., Pereira, F., & Sanchez, C. (2011). Design and properties of functional hybrid organic–inorganic membranes for fuel cells. *Chemical Society Reviews, 40* (2), 961–1005.
3. Shao, Y., Yin, G., & Gao, Y. (2007). Understanding and approaches for the durability issues of Pt-based catalysts for PEM fuel cell. *Journal of Power Sources, 171*(2), 558–566.
4. Cheng, S., Liu, H., & Logan, B. E. (2006). Power densities using different cathode catalysts (Pt and CoTMPP) and polymer binders (Nafion and PTFE) in single chamber microbial fuel cells. *Environmental Science & Technology, 40*(1), 364–369.
5. Sutharssan, T., Montalvao, D., Chen, Y. K., Wang, W. C., Pisac, C., & Elemara, H. (2017). A review on prognostics and health monitoring of proton exchange membrane fuel cell. *Renewable and Sustainable Energy Reviews, 75*, 440–450.
6. Chatenet, M., Dubau, L., Job, N., & Maillard, F. (2010). The (electro) catalystl membrane interface in the proton exchange membrane fuel cell: Similarities and differences with non-electrochemical catalytic membrane reactors. *Catalysis Today, 156*(3), 76–86.
7. Niu, X. D., Munekata, T., Hyodo, S. A., & Suga, K. (2007). An investigation of water-gas transport processes in the gas-diffusion-layer of a PEM FC by a multiphase multiple-relaxation-time lattice Boltzmann model. *Journal of Power Sources, 172*(2), 542–552.
8. Spurgeon, J. M., Walter, M. G., Zhou, J., Kohl, P. A., & Lewis, N. S. (2011). Electrical conductivity, ionic conductivity, optical absorption, and gas separation properties of ionically conductive polymer membranes embedded with Si microwire arrays. *Energy & Environmental Science, 4*(5), 1772–1780.
9. Ural, Z., Gencoglu, M.T., & Gumus, B. (2007). Dynamic simulation of a PEMFC system. In *Proceedings 2nd international hydrogen energy congress and exhibition IHEC*.

10. Offer, G. J., Mermelstein, J., Brightman, E., & Brandon, N. P. (2009). Thermodynamics and kinetics of the interaction of carbon and sulfur with solid oxide FCanodes. *Journal of the American Ceramic Society, 92*(4), 763–780.
11. Wang, L., Husar, A., Zhou, T., & Liu, H. (2003). A parametric study of PEM FC performances. *International Journal of Hydrogen Energy, 28*(11), 1263–1272.
12. AlZahrani, A., Dincer, I., & Li, X. (2015). A performance assessment study on solid oxide fuel cells for reduced operating temperatures. *International Journal of Hydrogen Energy, 40*(24), 7791–7797.
13. Jaouen, F., Goellner, V., Lefèvre, M., Herranz, J., Proietti, E., & Dodelet, J. P. (2013). Oxygen reduction activities compared in rotating-disk electrode and proton exchange membrane fuel cells for highly active Fe N C catalysts. *Electrochimica Acta, 87*, 619–628.
14. Al-Baghdadi, M. A. S. (2005). Modelling of proton exchange membrane fuel cell performance based on semi-empirical equations. *Renewable Energy, 30*(10), 1587–1599. Bevers, D., et al. (1997). Simulation of a polymer electrolyte FC electrode. *Journal of Applied Electrochemisrty, 27*, 1254–1264.
15. Kim, J. H., Jang, M. H., Choe, J. S., Kim, D. Y., Tak, Y. S., & Cho, B. H. (2011). An experimental analysis of the ripple current applied variable frequency characteristic in a polymer electrolyte membrane fuel cell. *Journal of Power Electronics, 11*(1), 82–89.
16. Wang, C. Y. (2004). Fundamental models for FC engineering. *Chemical Reviews, 104*(10), 4727–4766.
17. Berger, C. (1968). *Handbook of FC technology* (1st ed.). Englewood Cliffs: Prentice-Hall, Inc. 607.
18. Wandschneider, F. T., Finke, D., Grosjean, S., Fischer, P., Pinkwart, K., Tübke, J., & Nirschl, H. (2014). Model of a vanadium redox flow battery with an anion exchange membrane and a Larminie-correction. *Journal of Power Sources, 272*, 436–447.
19. Ercolino, G., Ashraf, M. A., Specchia, V., & Specchia, S. (2015). Performance evaluation and comparison of fuel processors integrated with PEM fuel cell based on steam or autothermal reforming and on CO preferential oxidation or selective methanation. *Applied Energy, 143*, 138–153.

Chapter 4
Failure Modes and Mechanisms

4.1 Overview

Fuel cell faults, transient or permanent, could be generally classified into electrical and chemical in nature. Depending on their severity, these faults may lead to a reduction in performance and, in advanced stages, may even destroy the cell. Failures that occur can be divided into three types:

1. Permanent faults (irreversible faults)
2. Transient faults (reversible faults)
3. External faults

4.2 Permanent Faults (Irreversible Faults)

This type of faults is permanent and, if continued, will lead to permanent damage to the fuel cell. The cell's performance and efficiency will degrade when such a fault occurs. The possible solution in this case is to replace the damaged fuel cell.

4.2.1 Membrane Degradation

Degradation of membranes is one of the most common faults that affect PEM fuel cell performance, efficiency and lifetime. This type of faults takes place over time such that the system's evolution is looking for new equilibrium points. In this case, the diffusion constants of the system change as well as pressure gradients between the two electrodes. Temperature, humidification and chemical attack cause changes in the characteristics of membrane and mechanical degradation. The chemical attack is caused by radicals initiating membrane degradation, because of their

A. Albarbar, M. Alrweq, *Proton Exchange Membrane Fuel Cells*,
https://doi.org/10.1007/978-3-319-70727-3_4

unpaired electrons being highly reactive. EIS is used to detect the membrane degradation; however, EIS is suitable only for a small number of cells (ten cells or less) due to the limitations of classical impedance metres. One of the most common factors that cause membrane degradation leading to reduced performance and reliability is chemical stability of the membrane. Because of their good features and properties, fluorinated compositions are relatively stabled in terms of chemical and thermal sense. For ion exchange, DuPont's Nafion and Gore's Primea membrane series are utilised. The lifetime of such membranes may be extended to thousands of hours. However, when temperatures are lowered to sub-zero, these membranes become more sensitive and vulnerable which may lead to rapid degradation. This happens especially in the case of transportation when the start-up and shutdown phases are under low temperatures [1].

4.2.2 *Absence of Catalyst*

This kind of fault occurs when the cell is flooded or when the fuel is contaminated with CO that lead to reduction in the active area of catalyst, which ultimately reduces the cell's performance. The output power could be used to diagnose the catalyst performance and is reversibly related to load resistance [2].

4.2.3 *Carbon Monoxide Effects*

Since the fuel cell's performance is directly affected by effectiveness of the used catalysts, therefore, when a saturated gas such as methane is used with hydrogen, carbon monoxide (CO) will be induced. This has a negative effect on cell's operation as platinum attracts CO into catalyst's surface. In addition, when voltage is between 5 and 6 volts (on the surface of the electrodes), hydrogen will react with oxygen and turn into CO_2, causing reduction in effective surface of the catalyst [3]. The CO atoms block membrane pores and prevent the crossing of ions – usually called 'poisoning the cells' – and lead to lower output.

This can be illustrated in the equations below:

$$CO + Pt \rightarrow Pt.CO \tag{4.1}$$

$$2CO + 2(Pt - H) \rightarrow 2(Pt.CO) + H_2 \tag{4.2}$$

The effects of CO on the performance of fuel cells may be reduced by the use of:

1. Platinum–ruthenium alloy (Pt–Ru) is added to the catalyst as well as platinum element. This reduces oxidation reaction voltage, on the surface of the catalysts, by 2–3 volts. This is explained in Eqs. 4.3, 4.4, and 4.5.

$$Pt + H_2O \rightarrow (Pt - OH) + H^+ + e^- \quad (4.3)$$
$$Ru + H_2O \rightarrow (Ru - OH) + H^+ + e^- \quad (4.4)$$
$$(pt = CO) + (Ru - OH) \rightarrow Pt + Ru + CO_2 + H^+ + e^- \quad (4.5)$$

2. Power electronic converters, connected to fuel cell terminals, to generate current pulses at low frequencies. These pulses raise anode voltages to levels adequate to convert CO into CO_2 by electro-oxidation process and free sufficient layers in PEM fuel cell catalyst.
3. High operating temperatures could help in preventing of CO poisoning. However, high temperatures also reduce the dynamics of cells.
4. Blowing a little air into the anode with fuel that may enhance oxygen availability to oxidise CO, which is absorbed in the layer of catalyst forming CO_2. This would help in cleaning Pt sites [3, 4].

Unfortunately, the above four solutions only mitigate the effects of CO poisoning on the performance of fuel cells rather than getting rid of this type of failures.

4.3 Transient Faults (Reversible Faults)

As the name implies, transient faults are reversible, so the fuel cell will go back to normal operation as soon as the fault is rectified. Usually this kind of faults is related to the operating conditions and the used control technique.

4.3.1 Hydration and Dehydration

Water behaviour and management are among the most important parameters influencing the performance and efficiency of PEM fuel cells. Therefore, optimisation of water management throughout cell's operation is recognised as critical, which is the subject of ongoing research works. Notably, water management needs to be considered in line with processes occurring in gas channels; across the GDL (gas diffusion layers), electrolytes and catalysts; and within the membrane to minimise the voltage losses. Although well-hydrated membranes are needed to ensure good performance, the gas channel and GDL must be also safeguarded from flooding [2]. The key phenomena and parameters that need to be considered and managed are RH (relative humidity) and the formation of water droplets in the gas channel, in addition to the water level in both the membrane and GDL [5]. Importantly, water is produced through electrochemical reaction, with the reactant gas that needs to be humidified to assist in membrane humidification and ensure ionic conductivity. In contrast, however, in-membrane water transport is led through back diffusion and electro-osmotic drag. Hence, ideal water levels are not easy to determine, and it is a difficult task to maintain, in particular, without in situ water-level measurement.

4.3.1.1 Fuel Cell Flooding

Due to electrochemical reactions, water is produced at the cathode side of the PEM fuel cells. During normal working operating conditions, water produced in fuel cells is used to humidify the membrane, to enhance its ionic conductivity, and the rest of water is removed by the airstream. If water production rate exceeds the removal rate, membrane pores and gas diffusion layers (GDL) will be blocked. Under this condition, cell's performance will degrade due to starvation of gas at both cathode and anode sides.

Losses in mass transportation are related to flooding conditions, especially at the cathode side, owing to a significant reduction in reactant transport rates to the electrocatalyst layers. Output voltage could be recovered relatively quickly as soon as purging of both sides of cathode and anode is completed. Table 4.1 shows the location and consequences of flood faults.

4.3.1.1.1 Cathode Flooding

Water flooding usually occurred at the cathode side based on the following three mechanisms:

1. Water is generated because of oxygen reduction reaction (ORR), and the amount of water continues to increase with the increase of current density and load.
2. Water particles, along with protons from anode to cathode, are pulled by ions, which is known as electro-osmotic drag phenomenon, under the effect of an electric field. The amount of water transferred is dependent on the membrane humidification level and may be increased by increasing the current density [6].
3. Flooding may also occur due to over-humidified reactant liquid water injection and gases.

Table 4.1 Location and consequences of flood faults

Fault location	The consequence of faults
Gas diffusion layer (GDL)	Loss of porosity and reduction in gas permeability
Gas channels	Low pressure in gas flow channels Low flow rate of gas channels Decay and change in gas channel geometry
Electrodes	Inhomogeneous pressures in bps Decay and change in flow-filled geometry
Membrane	Reduced mass transportation

The accumulated water on the cathode may be cleaned out from the porous electrode by using methods like water vapour diffusion or evaporation. Moreover, capillary transfer via GDL in the flow channels and back diffusion of water through membrane to anode may be used. However, the latter method is not so appropriate in the removal of water because back diffusion is only possible at lower current densities such as less than 0.3 A/cm^2 and at this current back diffusion can lead to transport of net water towards the anode [6], whereas the possibility of cathode flooding is more at higher current densities along with the possibility of excess of back diffusion due to electro-osmotic drag phenomenon [7]. It may result in increased content of water at the cathode ultimately increasing the cathode flooding. Evaporation and water vapour diffusion methods are usually used to remove water at a higher airflow rate with a higher temperature of cell operation. The process of flushing out the water from the cell is made possible when surface tension and viscosity increase [8]. However, the rise in temperature limits PEM fuel cell applications, especially in the transportation in addition to causing low durability of parts of the cell life, especially on the membrane.

High airflow rate may cause damage to delicate components such as the membrane. When it becomes impossible to efficiently remove the water which has been collected in the cathode, the water flooding, as presented in Fig. 5.2, affects FC performance badly. At higher current densities, such as above 0.55 A/cm^2, a considerable amount of cell voltage drop occurs due to reduction of partial-gas-pressure level at the cathode side because of flooding.

Moreover, by the drop-in cathode pressure from 1.5 kPa to 3 kPa, the initial potential (0.9 V) of the cell also drops to one third of the initial 0.9 V. Furthermore, water gathering and cathode flooding cause oxygen to enter the surface of the catalyst causing hindrance [9].

It is possible for the cathode voltage to drop significantly from 1.23 V to 0.00 V at current off and further from 0.8 to 0.1 V at current on with replacement of PRR with ORR. Current flows through the electrode along with the electrode polarisation, which causes a voltage shift in the negative direction; however, anode polarisation occurred towards the positive path. Anode potential tends to shift towards the positive direction with flowing current. Thereby, FC voltage may change from the initial 0.7 V to 0.2 V as shown in Fig. 5.2.

Cell voltage may be reversed from a positive to a negative value (i.e. from 0.7 V to −0.2 V) along with the replacement of PRR with ORR under the oxygen starvation phase. In PEM fuel cell, this process is known as the voltage reversal effect (VRE) [9]. However, when VRE occurs within a cell, it not only contributes to output of the stack but also may balance partial effective output voltage with single fuel cell (Fig. 4.1).

4.3.1.1.2 Anode Flooding

Water requires more time to gather at the anode side due to the production of water at the cathode side of the FC. Less flooding occurrence at the anode may cause

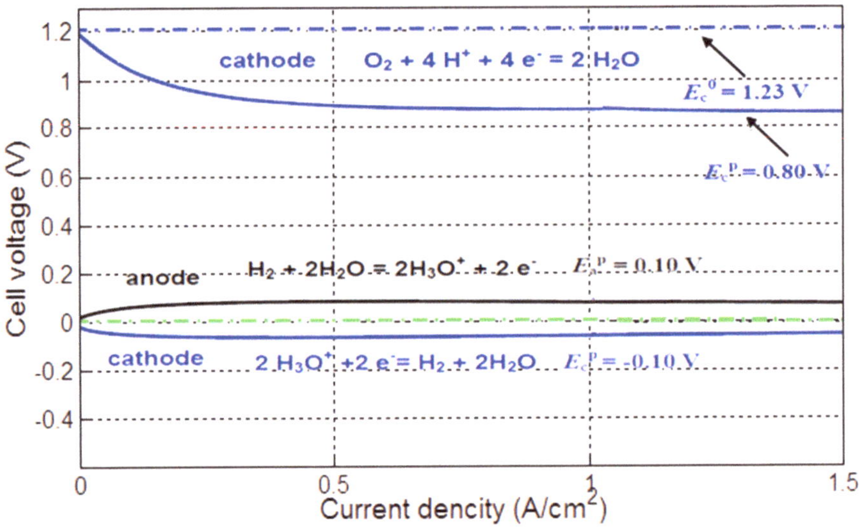

Fig. 4.1 The impact of voltage reversal in PEM fuel cell

serious concerns on the performance of the FC. Fault occurrences due to anode flooding may cause fuel starvation with carbon corrosion in layers of catalyst. Furthermore, at low flow rates of hydrogen, water may not be removed from the anode. Kim et al. [10] carried out experiments and found that water at the anode may be detected when the cell is operated at low current densities like 0.2 A/cm^2; however, electro-osmotic drag force may reduce the water content at the anode, especially at higher current densities. Moreover, anode flooding may also occur due to the lower temperature of the cell with higher condensation of water in channels [11].

Proton flux is high at the inlet of the anode where heavy electro-osmosis drags water molecules out of the anode to the cathode, which results in low water content. However, in contrast, at the outflow of the anode, current density is lower with higher water content.

Anode flooding may occur due to back diffusion of water from the cathode along with stream of fuel gas with a low hydration state. It is also possible that back diffusion of water, to anode through GDL, may surpass electro-osmotic effects in case the hydration state is not as high as that of the cathode. It was also observed in [10] by Kim et al. that water vapour which came from the GDL of the anode is condensed at the surface. However, water content due to condensation is not much in quantity. Injection of liquid water to cool and humidify is another reason for the occurrence of anode flooding [12].

4.3.1.1.3 Flow Channel Flooding

It is to be observed that the process of flooding may not only be related to porous electrodes of anode or cathode; however, it also occurs in the flow field gas channels. It usually depends on the interaction of the components' properties and the conditions of operations. Reactant gas opening may be hindered by the flow channel flooding and may weaken the power output of the FC. When serpentine channels are utilising flow channel flooding, one of the channels among them may cause starvation in some regions. Water flooding, along with loss of output voltages, may also cause changes such as:

1. Current density will be reduced, due to accumulation of water preventing the reactants from reaching active areas.
2. Released enthalpy of the condensed water may cause unbalanced local distribution of temperature and current density.
3. Gas flow resistance may rise due to accumulation of water in GDL; this may also result in an increased pressure between outlet and inlet of a FC [2, 12, 13].

4.3.1.1.4 Membrane Dehydration and Flooding

Dehydration and flooding of membranes are closely related to water management. Water management is a key factor in the occurrence of membrane dehydration or flooding due to a lack or/flooding of water. Although the membrane needs to be hydrated for good conductivity, flooding of the membrane will block the GDL.

The membrane dehydration usually occurred at the anode side. Poor water management is a key factor in the occurrence of membrane dehydration due to lack of water. When the cell works under dry run conditions of the membrane, it will lead to the deterioration of instantaneous and decline in the performance of the cell.

The low conductivity, due to the high-resistance cell, is linked to radical correlation relative humidity of the membrane. This will lead to a significant reduction in effort and electrical power in the cell. Although the temporary decline of the effort can be recovered by rehydrating the cell, operating the cell under dry conditions for a long time causes significant damage to cell which is irreversible, especially to the membrane. Recovery time is determined by membrane thickness and water diffusion. Due to higher water back diffusion that takes place at the anode in the bottom of the cell, the dehydration of the anode is more important at the inlet of the cell. Moreover, the membrane pores shrink because of the dehydration leading to lower back-diffusion rates. If the membrane worked for long periods under dehydrated conditions, the membrane becomes brittle, and it may develop cracks. In this case, the gas crossover causes an uncontrolled reaction for both H_2 and O_2, and hotspots will appear in the membrane leading to pinholes and more gas crossover. There are four reasons for cell dehydration:

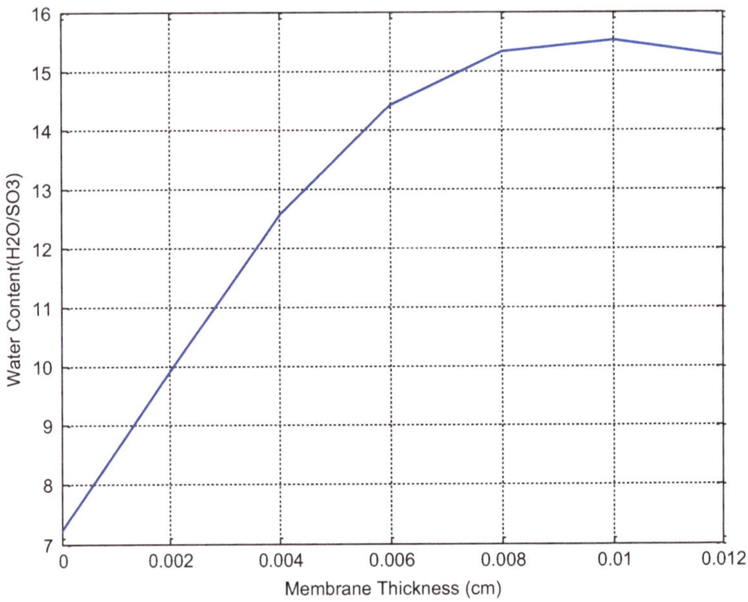

Fig. 4.2 Membrane thickness and water content

1. Inability to maintain adequate humidification of the cell when the cell is fed streams of dry gases or little humidification.
2. High temperature of the cell during operation will increase the water evaporation process.
3. Electro-osmosis can play a role in anode dehydration.
4. The membrane thickness.

Fig. 4.2 shows water content as a function in membrane thickness

4.4 Thermal Management and Its Impact on Performance

When a PEM fuel cell is exposed to freezing temperatures, thermal management becomes important. However, operation of a PEM fuel cell is still not completely understood in cases when it cycles below and above zero temperatures and cold starts. Though research has been carried out in recent years, yet more study is required to completely understand the challenges of operating under zero and elevated temperatures [14].

Fuel cell operation at temperatures higher than 100 °C has some advantages: increased tolerance for pollutants with improved efficiency and water management along with enhancement of the cooling mechanism because of higher temperature differences between coolant and cell. Low-quality hydrogen may be utilised due to reduction in CO poisoning and recovery of waste heat [15]. Degradation of electrode catalysts may also occur at moderate temperatures along with loss of its (Pt-) particles [14]. Chemical (in)stability is more important at higher temperatures. Redistribution and accumulation of particles are also accelerated at higher temperatures [16].

Another issue may arise at elevated temperatures when the potential is kept higher at the cathode. This may cause oxygen molecules to be split into atoms which may easily react with carbon or water, at high temperature, to form CO_2 and CO which may also result in carbon corrosion [14]. As a result, degradation of a catalyst may occur which ultimately affects the performance and lifetime. Moreover, at low relative humidity and elevated temperatures, water acceptance of the membrane worsens with the decrease in proton conductivity. This process may result in resistive loss, which may lower the cell efficiency and performance.

Song et al. [17] analysed the impact of Nafion contents on the performance of electrodes under three diverse temperatures, namely, 80, 100 and 120 °C. At 120 °C the effect of Nafion was greater than that at lower temperatures. Therefore, lower performance and mass transport problems of non-optimised Nafion content are observed.

4.4.1 Effect of Sub-zero Temperatures on Performance and Efficiency

The performance and efficiency of PEM fuel cell may be affected by exposing the non-operating PEM fuel cells to lower temperatures. The residual water may get frozen when the FC is exposed to sub-zero temperatures for long periods of time. This may result in mechanical damage or physical breakdown of the FC components due to mechanical and thermal stress. It is to be noted that the cause of degradation of the MEA is freezing water on Nafion membrane instead of inside the membrane. The volume of frozen water is expanded about 9% due to different densities of ice and water [18]. The process of ice formation and melting again in water can separate the catalyst layers from membrane layer of gas diffusion as well [19]. As the components are no longer in contact within the cell, mechanical damage may also cause a thermal and electrical contact as well.

Ge et al. [20] have linked the formation of ice at the surface of a membrane with its initial water content and initial thickness. Enclosed liquid water may flow out of the membrane and become frozen easily on surface because the Nafion membrane has a higher freezing temperature.

The thickness of the developed ice layer depends on the thickness of the membrane and higher content of initial water. For example, ice thickness for Nafion 112 is 5 mm, whereas for Nafion 117 it is 18 mm. Moreover, after several freezing cycles, fully hydrated membranes are cracked by carrying out tests by Plug Power

[21]. However, cracks may be less serious if after shutdown the hydration state of the membrane is lower.

Gas crossover due to cracks may lead to uncontrolled reaction between oxygen and hydrogen along with the formation of pinholes which ultimately damage the membrane and may cause a reduction in the lifetime of the cell [20].

Results in [22] show that cells which are being dried during the shutdown period may not experience physical damage and electrochemical losses.

4.5 Starvation

Starvation in PEM fuel cells occurs due to lack of fuel and/or air. This lack happens as a result of many issues such as faults in pump and compressor as well as the clog in pore layers due to the bad water management.

4.5.1 Fuel Starvation

Usually hydrogen is used as fuel, and oxygen is used as oxidant in PEM fuel cell. The starvation in hydrogen or oxygen will lead to a reduction in output voltage and thus affect FC reliability and performance. In the study performed by Gerard et al. [23], there was the analysis of performance degradation during air starvation of PEM fuel cell. Oxygen starvation occurs for several reasons, such as a fault during peak power demand, compressor delay or water management issues. In the same way, Liu et al. [24] investigated the effect of fuel starvation on PEM fuel cell performance, explored the changes in FC voltage and electrode voltage against the reversible hydrogen electrode (RHE) with time during air starvation and found inhomogeneous degradation of catalysts.

4.5.2 Air Starvation

Oxygen starvation occurs at the cathode due to lack of oxygen in the catalyst. Net flow rate of oxygen becomes equal to consumed oxygen in electrochemical reaction when steady-state conditions are achieved. However, in transient operating conditions, a sudden increase may occur for the power requirement of the FC. The system's air supply may lag behind the requirement causing lack of air and ultimately creating air starvation in the reaction, thereby increasing the concentration of oxygen at the cathode. However, if a shortage of oxygen occurs at cathode, proton (H^+) reduction reaction (PRR) may occur in place of ORR.

Dou et al. [25] analysed the current distribution during air starvation and observed the decreased current densities in the starvation region. In a similar way, Grigoriev et al. [26] introduced comprehensive understanding of the reversal process during air starvation. The current distribution measurement system and the local interfacial potential method are utilised in such a way as to find the correlation between the oxidant starvation and performance deterioration.

4.5.3 Reactant Leakage

A certain level of leakage rate must be maintained for hydrogen-fed PEM fuel cell because hydrogen gas has a higher combustibility level in small non-ventilated spaces. Critical hydrogen accumulation may occur due to an increase in leakage rate. This increment is because of cracks in the graphite plate along with seal ruptures and cross leaks, which may ultimately lead to explosions. It is argued by the author in [27] that to detect accurate hydrogen leakage, water vapour at the anode must be considered. However, a relative humidity sensor is also required, along with an already employed anode pressure sensor and mass flow sensor in the PEM fuel cell. Though it is expensive, it is deployed along with an adaptive alarm threshold in order to remove false alarms.

Miao et al. [27] intended to detect reactant leakage by observing the open-circuit cell voltages in aged cells in the region of the crossover and cathode/anode compartment. They analysed them at several different operating conditions, which ultimately helped to locate FCs with defects.

Signal-based fault detection is usually used for detecting anode/cathode leakage. After stopping the reactant flow, the amount of time needed is studied to reach a certain voltage threshold (0.5 V). It took a while for some cells to reach the set threshold; however, some cells reached their threshold very fast. To differentiate real faults, a normal probability curve is used. By studying the effect of correlation between water level in expanded vessels and reactant gas flow dynamics, anode/cathode compartment leakage is detected.

The PEM fuel cell simulator model was modified by Escobet et al. [28] which was designed by Pukrushpan et al. [29]. This was designed to include different faults such as an increment in compressor motor friction, compressor motor overheating because of water channel blockage, flooding in diffusion layer and cell temperature increase due to malfunctioning of the cooling system.

A knowledge-based fault monitoring system is proposed which is an interface engine with a user interface system conducted on LabVIEW. This is used for monitoring three fault modes:

1. Membrane drying
2. Air and hydrogen starvation of the electrochemical reaction
3. Leakage of the membrane

The research is based on the reality that membrane leakage is caused by the creation of holes in the membrane. Holes are formed due to hotspots, while another reason is fracture which is caused by mechanical stress. This stress is the result of cell operation under dynamic conditions of operation. This may result in a difference in pressure causing the membrane to be broken. To avoid leakage, a hydrogen pressure controller may be deployed.

4.6 Ageing and Degradation

In fuel cell (FC) technology, durability is considered as one of the major limiting factors. The lifetime of a FC is considered to be very short and after that cell performance degrades considerably over time. The operational time of the cell is estimated in [28] by utilising a pattern recognition-based approach, and the rest of the lifetime is calculated by using dynamic stack information, which is based on the EIS technique.

The process involved a two-step extraction method, which utilised an empirical model for extraction. This featured extraction is used on polarisation curves for the extraction of relevant descriptors while using a regression model. The imaginary part of the spectra is automatically split, in this model, into several parts which are estimated by polynomials. Real data sets are used for the evaluation of both approaches which estimate the lifetime of a fuel cell. Lifetime has a mean error of 24 h globally; however, the first method has 1000 h of mean error, the second method has 142 h of mean error and 95 h is the mean error by using featured extraction methods. Furthermore, the authors in [29] developed another method of pattern recognition, which is based on both dynamic and static stack information. This is based on EIS with a polarisation curve to enhance the estimation of operating time. This enhanced method is used to calculate the lifetime of a FC with a mean error of 2 h with 1000 h of global operating duration. Effect of harsh operating conditions on durability of PEM fuel cells is studied in [30]. It is done by feeding cell with low humidified air and dried hydrogen. This is done at temperatures greater than nominal temperature for 1000 h. Moreover, EIS and polarisation results are observed during tests conducted for several flow rates of gas. At 450 h, only a single cell has stopped the operation, whereas other two cells have relatively better performance and a low degradation rate.

In [31], an experiment is conducted to study the impact of ripple current with a high frequency of PEM fuel cell ageing. Modelled analysis proved that the current ripple does not impose many local conditions. However, analysis and results of the study are still questionable because a certain membrane model should be designed to see the real effect of ripple currents. Two protocols are designed in [32] by Zhang et al. that may be utilised to speed up the PEM fuel cell lifetime. One of the protocols is designed to operate at high temperature (300 °C).

Experiments showed that degradation in a stack's performance is found after 24 h of the experiment. The second protocol is designed to subject the stack to high current density (2 A/cm^2). Kim et al. conducted two ageing tests [33] and proved that the protocol can speed up the performance of the stack. The stack is operated under normal conditions in the first test. A constant current of 50 A is used for

1000 h which is being utilised with dry hydrogen. However, in the second test, dynamical load current is used for stack operation for 1000 h, whereas it is also being fed with dry hydrogen. Polarisation and internal resistance are utilised to evaluate the degradation in performance. Table 4.2 demonstrates important faults along with their causes and the effect of such faults on the performance and efficiency of PEM fuel cells.

Table 4.2 Common failure modes in PEM fuel cell

Modes of faults	Component of fuel cell	Types of faults	Fault consequences
Activation losses	Pt catalysts	Pt accumulation resulting in electrochemically active surface areas (EASA)	1. On/off load cycling repetition 2. Because of residual hydrogen, high voltage is induced at cathode equivalent 3. Carbon support loss
		Migration of Pt resulting in loss of catalyst materials	4. Platinum solubility during cell operation through air–hydrogen open circuit to air–air open circuit
		Loss of EASA due to absorption of atmospheric pollutants on platinum	5. Air impurities such as NO_2, H_2S and SO_2
		Loss of EASA due to absorption of fuel impurities on platinum	6. Hydrogen impurities such as CO and reduction of CO_2 to CO
		Catalyst losses because of chemical attacks and silicon/oxygen/platinum particle formation	7. Chemical degradation of silicone sealant
	Geometric structure of the catalyst layers	Increase in pore size and reduction in EASA due to deformation of catalyst structure by melting and freezing of water	8. Residual water in catalyst layers
	Carbon support in catalyst layer	FC is supplied with insufficient fuel and corrosion of carbon occurred to support current above provided hydrogen	9. Starvation of fuel 10. Partial coverage of hydrogen on catalyst layers 11. Localised flooding

(continued)

Table 4.2 (continued)

Modes of faults	Component of fuel cell	Types of faults	Fault consequences
Increase of mass transportation Losses	Porous and void regions of the FC, catalyst layer, GDL, BPP flow fields	Two-phase flow is induced by flooding caused by condensation of water	12. Low flow rate of channel gases 13. Low pressure in gas flow channels 14. Flow field geometry or configuration 15. Degradation of polymer electrolyte in catalyst layers due to impurity ions
		Hydrophobic material is used to treat loss of hydrophobicity in porous regions	16. Loss of material/material properties due to repeated thermal cycling/high operating temperature 17. Delamination caused by repeated thermal cycling 18. Exposure to sub-zero conditions damage material 19. PEM fuel cell overcompression, impacting the GDL
	Geometric structure of catalyst layer	Presence of isomer for impedance to transport attributable	20. Pores collapse due to ionomer expansion on water 21. Excessive ionomer loading 22. Formation of ionomer skins on catalyst layers 23. Lack of control during MEA processing
	Gas diffusion layer (GDL)	Ice formation with impedance to transport attributes	24. Presence of residual water from previous shutdown 25. Operation at sub-zero temperatures
		Loss of porosity, increased flooding and reduced gas permeability	26. BPPs transmitted inhomogeneous compression which are induced during operation

(continued)

Table 4.2 (continued)

Modes of faults	Component of fuel cell	Types of faults	Fault consequences
Increase of ohmic losses	Stainless-steel BPP	Loss of surface electrical Conductivity	27. Passivating layer on SS surface causes impedance to electron transfer
	Coated stainless-steel BPP	Loss of surface electrical conductivity	28. Coating defects such as pinholes and microcracks when coated with TiN 29. Degradation of conductive coating 30. Conductive polymer coating degradation
	Injection-moulded BPP	Low electrical conductivity	31. During injection-moulding process, polymer-rich boundaries are formed
	Polymer electrolyte membrane	Proton conductivity loss	32. Membrane hydration loss 33. Increase in water transfer coefficients and decrease in water diffusion due to replacement of protons 34. Decrease of water diffusion coefficient due to anisotropic swelling
Increase of efficiency loss, potentially leading to catastrophic cell failure	Polymer electrolyte membrane	Pinhole and crack formation	35. At sub-zero operating conditions, roughness of electrolyte membrane surface increases 36. Mechanical stresses induced by thermal hotspots in regions of strong electrical contact, i.e. BPP and high compression and reaction rates 37. Manufacturing process defects 38. Peroxide radical attacks of polymer end groups 39. Repetition of swelling and contraction because of hydrothermal stresses 40. Difference of excessive pressure between anode and cathode gas supply

(continued)

Table 4.2 (continued)

Modes of faults	Component of fuel cell	Types of faults	Fault consequences
Catastrophic cell failure	GDL/catalyst layer/membrane	Delamination of layers	41. Exposure and thermal cycling to sub-zero conditions
	Catalyst layer	Excessive adsorption of atmospheric and fuel impurities	42. Loss of tolerance to contaminants
	BPP	Cracking	43. Inhomogeneous compression 44. Mechanical shock/vibration 45. Irregularities in cell/stack construction
	Polymer electrolyte membrane	Gas leakage	47. Oxidation of seal polymer electrolyte membrane
		Short circuit	48. Formation of electrical network due to platinum distribution within self-humidifying membrane

4.7 Summary

Efficiency and performance of PEM fuel cells are affected by many internal and external factors. Degradation of materials, operational conditions and impurities or poisons are among those factors. Undoubtedly, while more conventional experimental approaches dedicated to cell state assessment (electrochemical impedance spectroscopy, current interruption, pressure drop and X-ray method) are undoubtedly powerful, they remain unsuitable in terms of handling the heterogeneous, transient and even stochastic elements of the phenomena witnessed in the membrane of the FC. Essentially, due to the cell geometry, chemical, electrical and physical value uniformity is not flawlessly achieved. In some regard, an approach facilitating membrane hydrogen state measurement, which needs to be sensitive to localised phenomena, would significantly contribute to developments made in water management.

References

1. Gittleman, S., Coms, F., & Lai, H. (2011). Membrane durability: Physical and chemical degradation. *Polymer Electrolyte Fuel Cell Degradation*, 15–88. ISBN 978-0-12-386936-4.
2. Alrweq, M., & Albarbar, A. (2016). Investigation into the characteristics of proton exchange membrane fuel cell-based power system. *IET Science, Measurement and Technology, 10*(3), 200–206. https://doi.org/10.1049/iet-smt.2015.0046. Online ISSN 1751-8830.
3. Ercolino, G., Ashraf, M. A., Specchia, V., & Specchia, S. (2015). Performance evaluation and comparison of fuel processors integrated with PEM fuel cell based on steam or autothermal reforming and on CO preferential oxidation or selective methanation. *Applied Energy, 143*, 138–153.
4. De Beer, C., Barendse, P. S., Pillay, P., Bullecks, B., & Rengaswamy, R. (2015). Electrical circuit analysis of CO poisoning in high-temperature PEMFC for fault diagnostics and mitigation. *IEEE Transactions on Industry Applications, 51*(1), 619–630.
5. Park, S., Lee, J. W., & Popov, B. N. (2012). A review of gas diffusion layer in PEM fuel cells: Materials and designs. *International Journal of Hydrogen Energy, 37*(7), 5850–5865.
6. Li, A., & Chan, S. H. (2013). Understanding the role of cathode structure and property on water management and electrochemical performance of a PEM fuel cell. *International Journal of Hydrogen Energy, 38*(27), 11988–11995.
7. Chen, L., Luan, H. B., He, Y. L., & Tao, W. Q. (2012). Pore-scale flow and mass transport in gas diffusion layer of proton exchange membrane fuel cell with interdigitated flow fields. *International Journal of Thermal Sciences, 51*, 132–144.
8. Speder, J., Zana, A., Spanos, I., Kirkensgaard, J. J., Mortensen, K., Hanzlik, M., & Arenz, M. (2014). Comparative degradation study of carbon supported proton exchange membrane fuel cell electrocatalysts–the influence of the platinum to carbon ratio on the degradation rate. *Journal of Power Sources, 261*, 14–22.
9. Therdthianwong, A., Saenwiset, P., & Therdthianwong, S. (2012). Cathode catalyst layer design for proton exchange membrane fuel cells. *Fuel, 91*(1), 192–199.
10. Kim, M., Jung, N., Eom, K., Yoo, S. J., Kim, J. Y., Jang, J. H., Kim, H. J., Hong, B. K., & Cho, E. (2014). Effects of anode flooding on the performance degradation of polymer electrolyte membrane fuel cells. *Journal of Power Sources, 266*, 332–340.
11. Yu, J. R., Matsuura, T., Yoshikawa, Y., Islam, M. N., & Hori, M. (2005). In situ analysis of performance degradation of a PEM FUEL CELL under non-saturated humidification. *Electrochemist. Solid-State Lett, 8*, 156–158.
12. Tian, G., Wasterlain, S., Endichi, I., Candusso, D., Harel, F., François, X., Péra, M. C., Hissel, D., & Kauffmann, J. M. (2008). Diagnosis methods dedicated to the localisation of failed cells within PEM FUEL CELL stacks. *Journal of Power Sources, 182*(2), 449–461.
13. Lee, D., & Bae, J. (2012). Visualization of flooding in a single cell and stacks by using a newly-designed transparent PEM FUEL CELL. *International Journal of Hydrogen Energy, 37*(1), 422–435.
14. Srouji, A. K., Zheng, L. J., Dross, R., Turhan, A., & Mench, M. M. (2013). Ultra-high current density water management in polymer electrolyte fuel cell with porous metallic flow field. *Journal of Power Sources, 239*, 433–442.
15. Karthikeyan, P., Li, H. C., Lipscomb, G., Neelakrishnan, S., Abby, J. G., & Anand, R. (2012). Experimental investigation of the water impact on performance of proton exchange membrane fuel cells (PEMFC) with porous and non-porous flow channels. In ASME 2012 international mechanical engineering congress and exposition. *American Society of Mechanical Engineers, 6*, 781–788.
16. Zhang, J. B., Kramer, D., Shimoi, R., Ono, Y., Lehmann, E., Wokaun, A., Shinohara, K., & Scherer, G. G. (2006). In situ diagnostic of two-phase flow phenomena in polymer electrolyte fuel cells by neutron imaging: Part B. Material variations. *Electrochimica Acta, 51*, 2715–2727.

17. Song, J. M., Cha, S. Y., & Lee, W. M. (2001). Optimal composition of polymer electrolyte fuel cell electrodes determined by the AC impedance method. *Journal of Power Sources, 94*(1), 78–84.
18. Jiao, K., & Li, X. (2011). Water transport in polymer electrolyte membrane fuel cells. *Progress in Energy and Combustion Science, 37*(3), 221–291.
19. Hou, J., Yu, H., Zhang, S., Sun, S., Wang, H., Yi, B., & Ming, P. (2006). Analysis of PEMFC freeze degradation at − 20 C after gas purging. *Journal of Power Sources, 162*(1), 513–520.
20. Ge, S., & Wang, C. Y. (2007). Characteristics of sub-zero start-up and water/ice formation on the catalyst layer in a polymer electrolyte fuel cell. *Electrochemical Acta, 52*(14), 4825–4835.
21. Theodorakakos, A., Ous, T., Gavaises, M., Nouri, J. M., Nikolopoulos, N., & Yanagihara, H. (2006). Dynamics of water droplets detached from porous surfaces of relevance to PEM fuel cells. *Journal of Colloid and Interface Science, 300*(2), 673–687.
22. Gebel, G., Diat, O., Escribano, S., & Mosdale, R. (2008). Water profile determination in a running PEMFC by small-angle neutron scattering. *Journal of Power Sources, 179*(1), 132–139.
23. Gerard, M., Poirot-Crouvezier, J. P., Hissel, D., & Pera, M. C. (2010). Oxygen starvation analysis during air feeding faults in PEMFC. *International Journal of Hydrogen Energy, 35* (22), 12295–12307.
24. Liu, Z., Yang, L., Mao, Z., Zhuge, W., Zhang, Y., & Wang, L. (2006). Behaviour of PEMFC in starvation. *Journal of Power Sources, 157*(1), 166–176.
25. Dou, M., Hou, M., Liang, D., Shen, Q., Zhang, H., Lu, W., Shao, Z., & Yi, B. (2011). Behaviour of proton exchange membrane fuel cells under oxidant starvation. *Journal of Power Sources, 196*(5), 2759–2762.
26. Grigoriev, S. A., Shtatniy, I. G., Millet, P., Porembsky, V. I., & Fateev, V. N. (2011). Description and characterization of an electrochemical hydrogen compressor/concentrator based on solid polymer electrolyte technology. *International Journal of Hydrogen Energy, 36*(6), 4148–4155.
27. Miao, Z., Yu, H., Song, W., Hao, L., Shao, Z., Shen, Q., Hou, J., & Yi, B. (2010). Characteristics of proton exchange membrane fuel cells cold start with silica in cathode catalyst layers. *International Journal of Hydrogen Energy, 35*(11), 5552–5557.
28. Escobet, T., Feroldi, D., De Lira, S., Puig, V., Quevedo, J., Riera, J., & Serra, M. (2009). Model-based fault diagnosis in PEM fuel cell systems. *Journal of Power Sources, 192*(1), 216–223.
29. Pukrushpan, J. T., Stefanopoulou, A. G., & Peng, H. (2004). Control of fuel cell breathing. *IEEE Control Systems, 24*(2), 30–46.
30. Lechartier, E., Laffly, E., Péra, M. C., Gouriveau, R., Hissel, D., & Zerhouni, N. (2015). Proton exchange membrane fuel cell behavioural model suitable for prognostics. *International Journal of Hydrogen Energy, 40*(26), 8384–8397.
31. Yu, X., & Ye, S. (2007). Recent advances in activity and durability enhancement of Pt/C catalytic cathode in PEMFC: Part II: Degradation mechanism and durability enhancement of carbon supported platinum catalyst. *Journal of Power Sources, 172*(1), 145–154.
32. Zhang, J., Xie, Z., Zhang, J., Tang, Y., Song, C., Navessin, T., Shi, Z., Song, D., Wang, H., Wilkinson, D. P., & Liu, Z. S. (2006). High temperature PEM fuel cells. *Journal of Power Sources, 160*(2), 872–891.
33. Kim, S. K., Choi, S. W., Jeon, W. S., Park, J. O., Ko, T., Chang, H., & Lee, J. C. (2012). Cross-linked benzoxazine–benzimidazole copolymer electrolyte membranes for fuel cells at elevated temperature. *Macromolecules, 45*(3), 1438–1446.

Chapter 5
Mathematical Modelling and Numerical Simulation

5.1 Overview

Fuel cells are complex devices, and their operation is associated with, simultaneously occurring, mechanical, physical and electrochemical phenomena. Clear understanding of these complex phenomena may enable effective improvement on the performance as well as the efficiency of the fuel cells, and hence manufacturing costs could be also brought down.

Realistic modelling and empirical verification methods are the most efficient way to study fuel cell device and the associated phenomena. The experimental method is better and more reliable, but cost and time constraints have made modelling a preferred option. The current density affects the fuel cell efficiency considerably, as explained in Chap. 3; thus it is important to measure its influence [1, 2]. However, experimental methods have also some limitations, e.g. it is not possible to measure the influence of current density on the efficiency of fuel cells under minor changes.

Again, although it is possible to obtain data and draw the V-I polarisation curve in the case of experimental techniques, it is extremely hard to discover the variables which influence the shape of the curve, especially when the FCs are working under various conditions. Along with this, the modelling techniques for FCs have a great advantage in that they can replicate the effects of a wide range of parameter values. This is highly beneficial since practical experiments can be conducted with a limited set of values at a single cell:

1. Electronic and protonic charge transfer.
2. Transfer of mass.
3. Electrochemistry of cell reactions.
4. Mechanical stress, i.e. compression of internal pressure.
5. Although it is a simple principle – the hydrogen and oxygen react with each other to produce electricity, heat and water – the modelling of cell phenomena is quite complex. In short, the phenomena of FCs are:

A. Albarbar, M. Alrweq, *Proton Exchange Membrane Fuel Cells*,
https://doi.org/10.1007/978-3-319-70727-3_5

(a) It occurs in the nanometre range and is modelled by the density functional theory and the principle of molecular dynamics.
(b) It occurs in the micrometre range, i.e. the material structure level.
(c) It also occurs in the millimetre to metre scale for the mass, flow dynamics, charge transfer and heat by the process of continuum mechanics.

The prime difference in time is due to the difference in speed between electrochemical reactions and the speed of transmission of heat and mass. Because of the overcomplicated nature of FCs, a single model cannot deal with all the possible phenomena that can occur within a FC. This is a general rule which is true for all modelling except for a few systems. The basic underlying physical principles of quantum mechanics are unable to exactly formulate the composition of the model at a macroscopic level. Even in the case of classical physics, the several phenomena which occur in a FC are translated to form a very complicated nonlinear differential equation-based system. In addition to this, the thermal expansion and the contact pressure distribution of the FC, along with the other mechanical stresses, are also responsible for the FC performance.

A FC model may have a time dimension to account for current drawn from the FC and changes in operating conditions, but the degradation of the cell must be accounted for. Even with modern computers, it is impossible to solve this system in three dimensions and under dynamic conditions along with handling different length scales ranging from nanosized catalyst particles to a macroscopic stack-sized structure. Due to this, any FC model always focuses on studying only a few effects in specific simple conditions. Typical examples of FC models used are electrochemical modelling of the FC and studying mass transport with the help of different flow field configurations. Early models were one-dimensional, even zero-dimensional and often studied only some of the parts of the FC (examples are the cathode electrode and the GDL), but due to the development of computers, three-dimensional models are now a common practice, and it is possible to model the whole geometry of the FC [3]. Even after this, many assumptions are still needed to remove complexity to reduce it to a solvable model. For example, sometimes the anode mass transfer is excluded in modelling due to limitations of performance. Another complexity is due to arbitrary parameter values being chosen due to the absence of experimental data or to the values being theoretically derived. For example, current density exchange is an important parameter for the electrochemical equations that comprise the basis of a FC model. Various factors affect this parameter, such as the amount of catalyst on the electrodes and the temperature of the cell. Hence, the values are different for each FC, and it becomes extremely difficult to measure its effects. It is common to simply use the parameter that fits with the model outcome in order to give a reasonable look and a physical feel.

Modelling faults, whether numerical errors or more profound problems, could have been hidden by the parameters chosen so that proper-looking results could be obtained. Hence, there is no guarantee that the given predictions for other parameters will be correct or not, i.e. the results cannot be predicted.

The exchange of current density is only one of the many examples; several other problematic parameters exist as in cases of modelling with two phases, like condensation and evaporation rates, and equations of capillary pressure [1]. It is usually also required to calculate averages for many material properties which are not isotropic such as porosity of GDL or the contact resistances between several layers in the cell, due to the absence of more detailed data. The results from these models can be accurate when considering the whole cell but can also be extremely inaccurate at explaining local phenomena.

5.2 Existing Modelling Techniques

A compact review of how FC modelling has evolved from the initial zero-dimensional analytic models to the complicated *3D* models that have been utilised so far will be provided in this section. The modelling studies discussed in this section have been classified for method, dimension and modelling aim. It is hard to categorise FC models, and there are many models, which do not fit into any of these categories, like spherical agglomeration modelling [3].

5.2.1 Analytical Models

One of the computational tools used by FC models is the finite element method (FEM). There are many models, which are based on analytically solvable equations. It is seen that the analytical type of FC model is always very simplified and idealised. The range of dimensions in which the model is solved is reduced to zero or one. The inherent physics is also simplified, usually by linearising the equations used, which cannot otherwise be solved analytically. The results are not very accurate, especially when the current density is so large. Analytical models can be used to gain approximate voltage–current dependencies and to perform short calculations on simple systems [3, 4]. In ideal conditions the basic view can be provided by analytical models. Following many early analytical models, 3D modelling is more common nowadays because of powerful computers but analytical, as well as semi-analytical, studies are performed even today.

5.2.1.1 One-Dimensional Models

1D Models can be analytical, but if the equations are not linearised, then they can be solved by using discrete and numerical algorithms. The benefit of this is that it makes the results more reliable. The current, voltage and other FC characteristics are the major limitations of 1D models.

Many local phenomena are beyond the range of 1D models because they assume unidirectional similarity in every direction; hence, the difference between the flow field plate ribs and the channels is not considered. The FC model represented in [5] is a 1D model with simple assumptions. The focus is mass transport, as well as management of water and cathode flooding.

5.2.1.2 Two-Dimensional Models

The types of 2D models found are channel models or cross-sectional models. The channel model may be used for studying reactant concentration, which varies along the channel due to consumption of the reactants. The cross-sectional models provide information regarding the process of electrical conduction and current in the flow plates. The 2D models assume an infinite planar plate. The flow field and 3D phenomenon cannot be studied using 2D models. The major components of 2D models are MEA and GDLs, and symmetry boundaries are used. The aims of a study make it possible to exclude some components; for example, if cathode mass transport is to be studied, then the anode side, as well as the membrane, can be excluded. The 2D versions of models are extremely useful in cases where the computational power is not large enough.

This sometimes happens if small geometrical details, like the thin layers of the electrode, are included as a domain of modelling. The mesh of the computational grid will be very fine in similar cases, thus making the model harder to solve. The 2D models can provide new local phenomena information [5]. These models are used very widely, as the 3D models can be either unnecessary or impractical for the study of FC phenomena. A few examples can be seen in, e.g. [5, 6]. The general form of 2D models is to focus on membrane transport phenomena, GDLs and local current distribution.

5.2.1.3 Three-Dimensional Models

Three-dimensional models can be considered as the most realistic as the FC geometry can be modelled exactly as it is, although several simplifications must be made. However, this means that, to solve these models, more computer capacity and time will be required. 3D models are best used while studying phenomena that cannot be modelled in 2D, like reactant flow and distribution phenomena. A 3D model can be used to obtain information about whether the reactants are evenly distributed or whether the distribution of current in the FC is uniform or not. Many early examples of 3D models, such as that presented in [7, 8], exclude MEA or the electrodes used, as they are very thinly layered and they greatly increase the computational grid size in a three-dimensional model. It is typical, rather than conducting studies of the whole FC, that only a section of the flow field is studied, such as one serpentine channel turn. There are many models which can cover the proper active area of a small-sized FC, such as the one given in [8]. In the diagram a

16 cm^2 FC was properly modelled to study the effects of many channel cross sections on the FC performance. It is very problematic that large-scale models are models of comparatively small-sized FCs, e.g. 49 cm^2 in [9].

Despite this application to small FCs, most real FCs are part of much larger sections to produce the required amount of power. The main problem is that, although the behaviour of the flow field is not harder to scale, the geometry of the channel works properly in small FCs, but they might not perform properly in the larger-sized FC due to changes in their Reynold's numbers. Hence, more effort is required to model the FCs when they are used in stacks, not just those which can be used in the laboratory. PEM fuel cell modelling has proved that the work carried out until now is important for good FC performance.

5.3 Semi-empirical Models

There are many conditions in which the FC phenomenon physics are not well understood or the correlations are inaccurate outside a certain parameter range, and failure to detect this may result in wrong conclusions. Due to this, the use of empirical correlations never helps in the purpose of understanding the physics of a FC. Hence, the use of empirical correlations may prevent us from finding fresh mechanisms, which could be manipulated to improve FC performance. Semi-empirical models can be accurate enough in design modelling and for materials that are in use already but can also predict new alternative materials or new innovative designs which could affect cell operation.

It is hard to model the phenomenon for practical modelling. Hence, researchers sometimes opt for formulas obtained from empirical calculations rather than those obtained from algebraic or differential calculations, which are more accurate mathematically in comparison to the former type. There are many cases in which the FC models have employed empirical correlations; an example of such a case is the partially humidified membrane conductivity calculations. In single FCs, semi-empirical modelling is more accepted. Semi-empirical models are used in stack modelling in which the models require more details and computing capacity.

The use of semi-empirical relations is necessary sometimes because of a lack of better alternatives, but whenever possible it must be avoided.

5.3.1 Dynamic Models

Models which include the time dimension are called dynamic models. In dynamic models FC operation is not constant but is dependent on time. Dynamic models are used for predictions of FC response or prediction for stacks, such as system-level changes or disturbances in the operating conditions. The start-up, shutdown and voltage cycling can also be studied with the help of this tool.

Dynamic models are extremely dependent upon empirical correlations, since including the time dimension requires significantly more computing capacity and time than those required for solving a steady-state model, for example, in which the FC may be modelled using an equivalent circuit model consisting of electrical components like resistors and capacitors which are connected to simulate a FC [10]. This makes the models much simpler. Hence, in several cases, the FC is not the object of study, because the constructed model focuses mainly on the system, as in [11], for example. Dynamic models using physical as well as electrochemical correlations can provide the required estimations of the reaction times of FCs reaching steady-state operation after changes in the operating conditions.

5.3.2 *Two-Phase Models*

One of the most important aspects of PEM fuel cells in terms of modelling, as well as FC operation, is mainly water management. Water is required to enhance the conductivity of ions in the membrane, but if too much water accumulates in the electrodes, then it will obstruct mass transport to and from the sites of reaction. Therefore, water management is very crucial for FC performance.

The two methods which are employed for the modelling of two-phase mass transport are the multiphase mixture-type model and the multiplied type model. In the multiphase mixture-type model, the phases are considered as a mixture for which equations can be solved [12, 13]. The information for each phase can be calculated from the solution in the mixture. In the multiplied model approach, both phases have a set of equations and both sets are solved simultaneously [14]. The latter method needs more computational capacity and a more efficient person, as convergence is comparatively hard to achieve. It also gives more accurate results and predictions for phenomena which cannot be achieved using the other method mentioned. Both models can only work with porous media flow, meaning cases in which Darcy's law is applicable, and thus it cannot be used for the purpose of modelling the behaviour of two-phase gas channels. This is considered problematic because the water accumulation process must be studied in three dimensions and across the field of flow and not just as a GDL and MEA cross section.

Some experimental techniques, like neutron imaging technology, electrochemical impedance spectroscopy (EIS) and X-rays, can be used to study the transport of water and its accumulation in the FC, but more accurate two-phase models are more useful for studying and improving PEM fuel cell water management. Several parameters for two-phase models are chosen in an arbitrary fashion or are derived from the results of experiments conducted on the coil [15].

It must be mentioned that several two-phase models sometimes use inaccurate boundary conditions for the channel boundary and the GDL where there is no liquid but the water is in gaseous form. This is clearly not correct as many experiments have shown the formation of water droplets in the channels [13]. Several studies have suggested that novel boundary conditions are the solution to this problem; for example, a basic model for droplet formation on the boundary can be used [14, 15]. However, the liquid water transported in the channels, which affects evaporation from the channels, has never been modelled correctly because Darcy's law does not apply in the channels, and hence the two-phase models are not modelled properly according to FC theory. After considering these issues, it is quite clear that two-phase modelling needs improvements before it can achieve trustworthy results.

5.4 Modelling Approach Proposed in This Book

The numerical modelling of the acoustic emission phenomenon, which takes place inside a FC, is a complicated process involving the coupling of nonlinear systems of equations. The acoustic emission phenomenon is dependent on other phenomena that require modelling first. These include species transport, heat transfer and water content.

The 3D and 2D finite element dependent commercial-type Multiphysics numerical equation solver (COMSOL 4.4) was used here to solve all the nonlinear transport coupled equations which describe the operations of AE.

By using the governing transport equation, the transportation of species is solved. The next step is to solve it using a loop-type algorithm. The PEM fuel cell computational domain is sectioned into a definite number of elements by using a second-order Lagrangian algorithm. The electrochemical engineering COMSOL module is used for this type of analysis, and the COMSOL physics application modes are used for solving the equations formed due to electrochemical phenomena. Many such equations are used, like the prime transport equations, the governing transport equations and the corresponding dependent variables:

1. The Navier–Stokes application mode, which is incompressible, for Newtonian-type reactant fluid flow behaviour simulation in the free media of the FC type domain.
2. The Navier–Stokes mode, which is incompressible, is coupled with Darcy's law, i.e. the Brinkman equation, for the simulation of the behaviour of the reactant fluid flow in porous media.
3. The Maxwell–Stefan multicomponent type diffusion and convection module for the transport of reactant species and products along the AE domain.
4. General heat transfer application mode in the FC domain by conduction and convection.

A stationary- and nonlinear-type numerical solver, combined with a direct-type linear solver (UMFPACK), is used because the source charge, in terms of conservation, makes it a nonlinear problem. The comparative tolerance for the error was set at $1\text{x}10^{-4}$. In the case of parametric analysis, the parametric-type solver of COMSOL is used here, and the parametric variables are defined with a specific analytical range. The performance polarisation curves were obtained by varying the FC potential using the parametric solver of COMSOL. The parametric solver uses an iteration loop to change the variable values in the specified range.

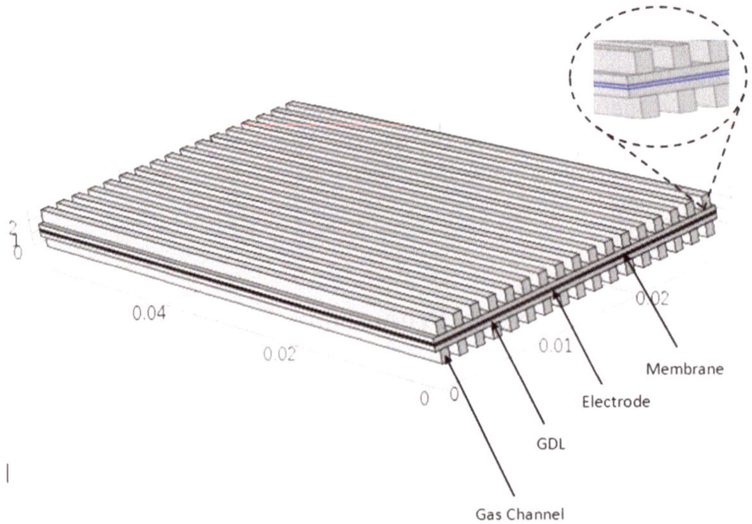

Fig. 5.1 Three-dimensional diagram of the proton-exchange membrane fuel cell

The convergent behaviour of this nonlinear-type problem is very sensitive to the first estimation of the given solution. Hence, to accelerate the process of the numerical convergence of the given nonlinear-type problem, the incompressible type of Navier–Stokes equations, along with the Brinkman equation, is first solved together to obtain a better estimation of the convective-type viscous properties of the flow of the reactant. After this, the fully coupled type, which involves all the types of physics modes, is solved by using these values for an initial guess. This technique is used to avoid possible numerical instability in the convective-type terms in small-sized channels of the gas flow distributor. This is because the numerical modelling of flow behaviour in such small-sized channels can ultimately lead to fluctuations at the walls, which can further lead to numerical instabilities and may result in the simulation crashing. Hence, to minimise the numerical instabilities in such small flow channels, the useful concept of an artificial kind of diffusion is used in this case. Streamline diffusion is used here with a value tuning parameter of 0.25 and introduces stabilising terms to the Navier–Stokes equations of incompressible type to reduce the numerical instabilities in the given streamline direction for the fluid flow; see Fig. 5.1.

5.4.1 Modelling Procedure

This section gives a general view of the underlying physics used in the models employed – see details in Chap. 7. Modelled phenomena consist of transfer of mass and heat, along with charge, as well as electrochemical reactions and ultimately the emitted acoustic signal (acoustic emissions – AE).

These are modelled with proper partial differential equations, as well as with boundary conditions which depend on the given model.

5.4.1.1 Assumptions

The assumptions which are made are as follows:

1. Heat transfer is negligible and is achieved by conduction in the gas phase.
2. The type of electrode layer used is 'ultrathin'; hence, the diffusion of the gas in the electrode player can be easily neglected.
3. The gas mixture used here behaves ideally.
4. Liquid water is available in the form of small droplets.
5. The electro-osmotic type of coefficient, along with the coefficient of diffusion in the membrane for water, is mainly determined by the water activity in the channel of the anode flow. This is especially valid at places with high current densities because at this stage it is most likely that the membrane's anode side is drier in comparison with the cathode side.

Tables (5.1) and (5.2) illustrate the operating conditions and geometric properties of the PEM fuel cell computational domain that is used in this book.

5.4.1.2 Mass Transfer

Models of the reactant's transport are the building blocks of mass transfer modelling, which consists of the inert substances along with the reaction products in the

Table 5.1 Geometric properties of the PEM fuel cell computational domain

Parameter	Parameter	Parameter
Gas channel length	0.05	m
Gas channel width	0.05	m
Gas channel height	0.005	m
Perforated sheet thickness	0.00055	m
Perforated hole diameter	0.003	m
GDL thickness	0.0002	m
Membrane thickness	0.00018	m

Table 5.2 Operating conditions value

Parameters	Value	Unit
Oxygen inlet mass fraction	0.226	
Water inlet mass fraction	0.0934	
Hydrogen inlet mass fraction	0.844	
Operating temperature	298–353	K
Relative humidity, air & H2	70–99%	

FC. In PEM fuel cells, the reactants are H_2 on the anode and O_2 on the cathode, and the reaction product is water. The inert phase is N_2 or a mixture of N_2 and A. Both of these gases are very much present in air and hence in the cathode part of a FC which operates in air. They never participate in the reactions, but they do affect the transport of mass inside the body of the FC. Mass transport in the anode is excluded from the models used, as the phenomenon of performance limiting usually occurs on the cathode side because of the slower type of reaction kinetics mechanism, which increases the risk of flooding. In this book, the distribution of electrode reactants is the focus of mass transport modelling, on the one hand is the flow field study on the scale of the entire FC and on the other hand is the study of the scale of individual ribs and channels.

Species are transported on to the cathode through convection as well as the diffusion method. A pressure difference of an external nature is applied across the FC creating the FC flow. In the free convection case, density variations are produced in the air, caused by differences in concentration and temperature. The reactants are consumed on the electrode, and reaction products are generated on the electrode as well, which creates a density gradient and concentration gradients. The fluid flow passes into the channels, the electrode and the diffusion layer of porous gas. The membrane can transport water, but it is impermeable to gases.

The fluid flow is modelled in the channels, with the continuity equations and the Navier–Stokes (4.1) and (4.2). These equations model standard laminar equations for flows of an incompressible nature. There is no exact change limit for laminar to turbulent flow, but flow of the fluid can usually be laminar if the flow Reynolds number is below 2300. In the FC-type environment, this condition is actually met, and the flows are relatively slower. In the gas diffusion layer as well as the electrode, Darcy's law is used.

$$-\rho u.\nabla + \rho g - \nabla P + \nabla(\varepsilon(\nabla u + \nabla uT)) = S\_(SN) \tag{5.1}$$

$$\nabla.(\rho u) = S_{\text{con}} \tag{5.2}$$

$$\nabla.\left(\rho.\frac{\varepsilon}{k}\nabla p\right) = S_{\text{D}} \tag{5.3}$$

where:

ρ: the average fluid density
u: the velocity vector
P: the pressure
k: the permeability of the porous medium
S: the source of the term of every equation
ε: the average viscosity of the fluid

Equations (5.2) and (5.3) model the velocity pressure, in terms of fluid behaviour in the frame centre mass. For driven flow concentration calculations, diffusion should also be modelled. Using the simple diffusion, equation (i.e. Fick's law) in a FC cathode is considered incorrect, however, as the fluid has three constituents (oxygen, water and nitrogen) that have different sources, diffusion and sinks constants. Hence, diffusion equations of the multicomponent type or of the Maxwell–Stefan type, (5.4) and (5.5), have to be employed.

Although the Maxwell–Stefan equations for each component can be written, they need to be solved only for the first two, as the third component can be calculated from the first two, as the mass fractions will add up to one. Hence, the mass fraction of nitrogen (N_2) is calculated from those of water (H_2O) and oxygen (O_2).

$$\nabla.\left(-\rho\omega_i \sum_{j=O_2,\,H_2O,\,N_2} \widetilde{D_{ij}}\left(\nabla x_j\right) + \rho\omega_i u\right) = S_i^{\text{mass}} \tag{5.4}$$

Here $i = O_2, H_2O, N_2$

$$\nabla x_j = \frac{M^2}{M_j} \sum_{\substack{K = O_2, H_2O, N_2 \\ K \neq j}} \left(\frac{1}{M} + \omega_k\left(\frac{1}{M_K} - \frac{1}{M_j}\right)\right)\nabla\omega_k \tag{5.5}$$

Here, ω_i is the species mass fraction of i, $\widetilde{D_{ij}}$ represents the Maxwell–Stefan diffusion coefficient, M_j represents the molar mass and x_j the species molar fraction of j. Maxwell–Stefan equations can be expressed for molar fractions as follows:

$$\nabla.(Cx_i u) - \nabla.\left(C\widetilde{D_{\text{eff}}}\nabla x_i\right) = S_i^{\text{molar}} \tag{5.6}$$

The formulations in Eqs. (5.4) and (5.6) are similar and both forms are used in this book.

The source term, S, used in Eqs. (5.1), (5.2) and (5.3) arises from the reactions in the PEM fuel cell. They consume oxygen (O_2) and produce water (H_2O). Hence, they only exist in the electrodes and not anywhere else. The exceptions are the source terms in the Navier–Stokes continuity equations in that they are used in the model of the flow field in which the only PEM fuel cell component modelled is the flow field and hence the reaction's effects are modelled as if they have occurred inside the channels.

$$S_{\text{NS}} = -\frac{i_o A}{2FV_{\text{Ch}}} M_{o_2} u \tag{5.7}$$

$$S_{\text{p}} = -\frac{i_o}{4F}\left(M_{\text{Ho}_2} - M\text{o}_2\right) \tag{5.8}$$

$$S_{\text{Con}} = -\frac{i_o A}{zFV_{\text{Ch}}}\left(\xi M_{\text{Ho}_2} - M_{o_2}\right) \tag{5.9}$$

$$S_i^{\text{Mass}} = \begin{cases} -\frac{i_o}{4F} M_{\text{O}_2} & , i = \text{O}_2 \\ \frac{i_o}{2F} M_{\text{HO}_2} & , i = \text{H}_2\text{O} \end{cases} \tag{5.10}$$

$$S_i^{\text{Molar}} = \begin{cases} -\frac{i}{4F} & , i = \text{O}_2 \\ \frac{i}{2F} & , i = \text{H}_2\text{O} \end{cases} \tag{5.11}$$

where i_o represents the electrode current density reactant, A is the active area, z is the number of electrons in the reaction, F is Faraday's number and ξ is the portion of water in the reaction products leaving the FC from the anode, while V_{Ch} is the channel's volume. A part of the reaction's water product can be from the anode to the cathode driven by methods of diffusion.

The local current density changes along the channel length as the overvoltage of the electrodes as well as the conductivity of the membrane varies. The parameter signifies the net water molecules per unit of proton flux ratio [16]. The method of calculation is as follows:

$$\text{Net water flux} = N_{\text{w,y,a}}(x) = \alpha \frac{I_0}{F} = n_{\text{d}} \frac{I_o}{F} - D_{\text{w}} \frac{dc_{\text{w}}}{dy} \tag{5.12}$$

where the effect of migration is the first term on the right side and the second represents diffusion. Manipulation of this yields the following equation:

$$\alpha = n_{\text{d}} - \frac{F}{I_0} D_{\text{w}} \frac{dc_{\text{w}}}{dy} \tag{5.13}$$

This can be simplified by assuming the difference in water accumulation of the cathode with respect to the anode that can be approximated by a linear difference of single step.

$$\alpha = n_{\text{d}} - \frac{F}{I_{\text{a}}} D_{\text{w}} \frac{c_{\text{w,c}} - c_{\text{w,a}}}{t_{\text{m}}} \tag{5.14}$$

where t_{m} is the thickness of the membrane. The parameter 'n_{d}' is the electro-osmotic drag coefficient, which is the same as the number of water molecules carried by a proton. This quantity is highly dependent on the content of the water in the given membrane, which is a function of the water activity in the gas phase adjacent to that membrane. The partial dehydration that exists along the anode, as well as the saturation which exists along the cathode, is very likely to occur especially at current densities of higher magnitude. This is due to the higher transport rate of water because of the electro-osmosis (also called drag) from the anode to cathode compared to the back-diffusion rate of the water from the cathode and the anode. Physically, this means that the water content of the anode side is lower. For the same reason, the water activity on the anode side may be used to calculate the electro-osmotic coefficient in the membrane. The electro-osmotic coefficient is a function of the water activity in the low channel of the anode used [17] and can be expressed as:

$$n_{\mathrm{d}}=\begin{cases}0.0049+2.02a_{\mathrm{a}}-4.53a_{\mathrm{a}}^{2}+4.09a_{\mathrm{a}}^{3} & a_{\mathrm{a}}\leq 1\\ 1.5+0.159(a_{\mathrm{a}}-1) & a_{\mathrm{a}}>1\end{cases} \tag{5.15}$$

The parameter, D_{w}, needs to be calculated. This gives the coefficient of water diffusion in the membrane. It is believed that the coefficient of diffusion of water in a given membrane is dependent on the content of the water in the membrane, in the same way as the coefficient of electro-osmosis.

$$D_{\mathrm{w}}=n_{\mathrm{d}}D^{o}\exp\left[2416\left(\frac{1}{303}-\frac{1}{T_{\mathrm{s}}}\right)\right] \tag{5.16}$$

The water accumulation is a function of the water activity in the electrodes and is given by:

$$C_{\mathrm{w,k}}=\begin{cases}\dfrac{\rho_{\mathrm{m,dry}}}{M_{,\mathrm{m,dry}}}(0.043+17.8a_{k}-39.8a_{a}^{2}+36.0a_{ak}^{3} & a_{\mathrm{k}}\leq 1\\ \dfrac{\rho_{\mathrm{m,dry}}}{M_{,\mathrm{m,dry}}}(14+1.4(a_{\mathrm{k}}-1)\,) & a_{\mathrm{k}}>1\end{cases} \tag{5.17}$$

Either the anode or the cathode is substituted for the subscript k. $\rho_{\mathrm{m,dry}}$ and $M_{,\mathrm{m,dry}}$ are the dry proton-exchange density and the equivalent weight of a. The anode and cathode water activity are as follows:

$$a_{\mathrm{a}}=\frac{x_{\mathrm{w,a}}P}{P_{\mathrm{w,a}}^{\mathrm{sat}}}=\left(\frac{M_{\mathrm{w,a}}^{\mathrm{v}}}{M_{\mathrm{w,a}}^{\mathrm{v}}+M_{\mathrm{H_2}}}\right)*\frac{P}{P_{\mathrm{w,a}}^{\mathrm{sat}}} \tag{5.18}$$

$$a_{\mathrm{c}}=\frac{x_{\mathrm{w,c}}P}{P_{\mathrm{w,c}}^{\mathrm{sat}}}=\left(\frac{M_{\mathrm{w,c}}^{\mathrm{v}}}{M_{\mathrm{w,c}}^{\mathrm{v}}+M_{\mathrm{O_2}}+M_{\mathrm{N_2}}}\right)*\frac{P}{P_{\mathrm{w,c}}^{\mathrm{sat}}} \tag{5.19}$$

The water vapour can be determined by the following equation:

$$\begin{aligned}\log_{10}P_{\mathrm{w,k}}^{\mathrm{sat}}&=2.95*10^{-2}(T_{\mathrm{k}}-273.15)-9.18*10^{-5}*(T_{\mathrm{k}}-273.15)^{2}\\&\quad+1.44*10^{-7}*(T_{\mathrm{k}}-273.15)^{3}-2.18\end{aligned} \tag{5.20}$$

The change in mole number of every reactant can easily be determined by performing a substitution of Eqs. (5.13, 5.14, and 5.15) into Eq. (5.18). Equation (5.19) describes the methods by which the reactants can be consumed:

$$\frac{dM_{i}}{dx}=-hN_{(\mathrm{H_2,O_2,N_2}),\mathrm{y,k}}(x) \tag{5.21}$$

The change in the number of moles of liquid water in each flow type channel is determined by the rates of evaporation and condensation [18]:

$$\frac{dM_{\mathrm{w,k}}^{\mathrm{l}}}{dx} = \left(\frac{K_{\mathrm{c}}hd}{RT_{\mathrm{k}}}\right) * \left(\frac{M_{\mathrm{w,k}}^{\mathrm{v}}}{M_{\mathrm{w,k}}^{\mathrm{v}} + M_{\frac{\mathrm{H_2}}{\mathrm{O_2}}}} * P - P_{\mathrm{w,k}}^{\mathrm{sat}}\right) \tag{5.22}$$

where k_{c} is the homogeneous rate constant for condensation as well as for water evaporation, while h and d are the width and height of the given channel. The amount of liquid water in the flow channels is directly proportional to the difference between the partial pressure and the water vapour pressure. Physically, this simply means that liquid H_2O will condense if the partial pressure of water vapour becomes higher than the vapour pressure, while if liquid water is also present and the partial water vapour pressure is lower than its vapour pressure, then the liquid H_2O will surely vaporise. The change in the number of moles of water vapour along the flow channels is given by the following equation:

$$\frac{dM_{\mathrm{w}}^{\mathrm{v}},\mathrm{k}}{dx} = -\frac{dM_{\mathrm{w}}^{\mathrm{l}},\mathrm{k}}{dx} - hN_{\mathrm{w,y,k}}^{\mathrm{v}}(x) \tag{5.23}$$

The factors mentioned above affect the amount of water vapour in the flow channels:

1. The reaction of oxygen (O_2) with the proton and electron at the cathode produces H_2O.
2. The water (H_2O) generated at the cathode may be used in the membrane up to the anode because of the difference in concentration.
3. Here, the protons which migrate from the anode up to the cathode can bring the water vapour with them (i.e. via drag or electro-osmosis).
4. Water vapour condensation and liquid water evaporation are dependent on the difference between partial pressure and vapour pressure.

The first term on the right-hand side of Eq. (5.23) signifies condensation as well as evaporation, while the second term is the total water vapour transport across the membrane. The water transport is the result of the difference in concentration, pressure and number of water molecules, which are carried by the migrating protons.

5.4.1.3 Charge Transfer

The transfer of charge in a PEM fuel cell consists of the movement of electrons and ions. Hence, it is important to model both ionic and electronic potential individually [19].

The former exists in the membrane and electrodes only and the latter in the electrodes and GDLs only. Charge transfer, of ionic or electronic nature, is modelled by:

$$-\nabla.(\sigma_{\text{membrane,GDL}} \nabla \psi_{\text{m}}) = S_{\text{m}} \tag{5.24}$$

The conductivity, $\sigma_{\text{membrane, GDL}}$, is a membrane humidity function. The source terms S_i, representing ionic potential, and S_e, representing electronic potential, are positive in the electrodes:

$$S_{\text{m}} \begin{cases} i_{\text{a}} \\ -i_{\text{c}} \end{cases} \quad \text{and} \quad \begin{cases} -i_{\text{a}} \\ i_{\text{c}} \end{cases} \tag{5.25}$$

The reaction current densities, i_{a} and i_{c}, at the electrodes are mathematically calculated as per Eq. (3.18) in Chap. 3. An aspect of heat and charge transfer in a FC is that the variables are not continuous in different FC components. This is because of the components' contact resistances, like that between the GDL and the electrodes. There is spatial variation in contact resistances, as compression of the components produces such changes on the scale of the whole active area as well as locally in the ribs and the channels [20]. The differences in thermal and electric contact resistances have a large effect on FC performance and lifetime.

5.4.1.4 Heat Transfer

Compared to FCs such as SOFCs, PEM fuel cells operate at a much lower temperature, but transfer of heat is an essential factor for the performance of FCs. Due to local temperature variations, as well as in the current density, hot spots occur. The FC performance is not affected significantly by hot spots, but they can cause premature degradation of the membrane, and hence they can shorten FC lifetime. Modelling of heat transfer is very straightforward. There are only two methods of transfer of heat inside the FC, namely:

1. Convection
2. Conduction

Heat transfer by radiation is not very significant in PEM fuel cells as the temperature and the contact resistances and the thermal bulk are comparatively low. Heat transfer is modelled according to Eq. (5.26):

$$\nabla.(k\nabla T) = \sum_i C_{p,i}, i\rho_i u.\nabla T + S_{\text{T}} \tag{5.26}$$

Heat is produced from both reactions and by ohmic heating due to the electronic and ionic currents. Hence, the thermal source, S_{T} ,in each region is:

$$S_{\mathrm{T}}\begin{cases} \sigma^{\mathrm{s}}{}_{\mathrm{GDL},}\left(\nabla\psi_{\mathrm{G}}\right)^{2} \\ \sigma^{\mathrm{m}}{}_{\mathrm{membrane},}\left(\nabla\psi_{\mathrm{m}}\right)^{2} \\ \sigma^{\mathrm{s}}{}_{\mathrm{CL},}\left(\nabla\psi_{\mathrm{s}}\right)^{2}+\sigma^{\mathrm{m}}{}_{\mathrm{CL},}\left(\nabla\psi_{\mathrm{m}}\right)^{2}+i_{\mathrm{a}}\phi_{\mathrm{a}}+\dfrac{i_{\mathrm{a}}T\Delta S_{\mathrm{a}}}{2*F} \\ \sigma^{\mathrm{s}}{}_{\mathrm{CL},}\left(\nabla\psi_{\mathrm{s}}\right)^{2}+\sigma^{\mathrm{m}}{}_{\mathrm{CL},}\left(\nabla\psi_{\mathrm{m}}\right)^{2}-i_{\mathrm{c}}\phi_{\mathrm{c}}-\dfrac{i_{\mathrm{c}}T\Delta S_{\mathrm{c}}}{4*F} \end{cases} \tag{5.27}$$

Here, σ is the ionic or the electronic conductivity of the GDL for catalyst and membrane (CL), ψ_{i} and ψ_{e} are the electronic and ionic voltages, ϕ_{a} and ϕ_{c} are the overvoltage of the cathode and anode, i_{a} and i_{c} are the densities of the reaction current at the cathode and anode, T is the temperature and ΔS is the change in reaction entropy [21]. From the given Eq. (5.26), we can note that as i_{c} is of negative magnitude, the cathode source terms have negative signs.

The transfer of charge in a PEM fuel cell is constituted by the movement of electrons and ions. Hence, it is required to model both electronic and ionic potentials individually.

The former exists only in the electrodes and membrane, while the latter exists in the GDLs and electrodes. The charge transfer, whether ionic or electronic, is modelled by:

$$-\nabla.\left(\sigma_{\mathrm{membrane,GDL}}\nabla\psi_{\mathrm{i,e}}\right)=S_{\mathrm{i,e}} \tag{5.28}$$

The conductivity, $\sigma_{\mathrm{membrane,GDL}}$, is a membrane humidity function. The source term S_{i} represents the ionic potential, while S_{e} represents the electronic potential, and both are non-zero in the electrodes:

$$S_{\mathrm{m}}=\begin{cases} i_{\mathrm{a}}, & \text{at anode} \\ -i_{\mathrm{c}}, & \text{at cathode} \end{cases} \quad \text{and} \quad S_{\mathrm{s}}=\begin{cases} -i_{\mathrm{a}}, & \text{at anode} \\ i_{\mathrm{c}}, & \text{at cathode} \end{cases} \tag{5.29}$$

The reaction current densities, i_{a} and i_{c}, are calculated at the electrodes according to Eq. (3.18) given in Chap. 3. An aspect of charge and heat transfer in FCs is that these variables are noncontinuous between different components of the FC [22, 23]. This is because of the contact resistances between the FC components, e.g. the electrodes and the *GDL*. Spatial variation is present in contact resistances, as the compression applied to these components varies on the local scale of the ribs and channels and on the scale of the completely active area. The differences in electric and thermal contact resistances have an important effect on FC performance and lifetime.

5.4.2 A Novel Semi-empirical Model

In addition to 3D and 2D models, realistic mathematical models were developed and implemented in a MatLab/Simulink® environment to get an overview of the system's behaviour. Some parameters, such as temperature, pressures, water content and current density, have been studied to evaluate the effect of PEM fuel cell system reliability. The proposed model depends on principle equations, combined with equations having experimentally defined parameters, thus resulting in a semi-empirical system. The mathematical model equations describe the operation of the FC composed of the voltage–current characteristics and a relationship for the temperature and pressure as a function of the current density from the FC.

As mentioned earlier in Chap. 3, there are three different types of losses that occur within the FC, namely, activation losses, ohmic losses and concentration losses.

5.4.2.1 Modelling of Activation Polarization

The activation losses have been considered in two parts: activation losses 1 (Act 1) and activation losses 2 (Act 2). Activation losses 1 occurs due to the temperature and current-dependent voltage drop. However, it has been converted into equation univalent resistance first, and then it will be combined with concentration losses and the capacitor due to the double-layer changing effect to give the voltage drop (Act 2). The activation polarization was modelled based on the equation below:

$$\eta_{\text{act}} = -0.9514 + T*0.00312 + T*7.4*10^{-5}*\left[\ln\left(c*\text{o}_2\right]\right. - 0.000187*\ln\left(i_{\text{o}}\right) \tag{5.30}$$

$$C*\text{O}_2 = \frac{\text{P}_{\text{O2}}}{5.08*10^6\text{EXP}\left(\frac{-498}{T}\right)} \tag{5.31}$$

where i_o is the density of current and $C*O_2$ is the density of the hydrogen on the catalyst layer (Figs. 5.2, 5.3, 5.4, 5.5 and 5.6).

5.4.2.2 Modelling of Ohmic Polarization

The ohmic polarization (η_{ohm}) is positively proportional to the current. Since the cell's resistance is almost constant, the ohmic polarization changes linearly, and this is because of the emergence of resistance while crossing of ions in the electrolyte and ohmic resistance electrodes. The reduction of these resistances by using the appropriate electrolyte and metals in the electrodes can be adopted to overcome this problem, and the equation can be written as follows:

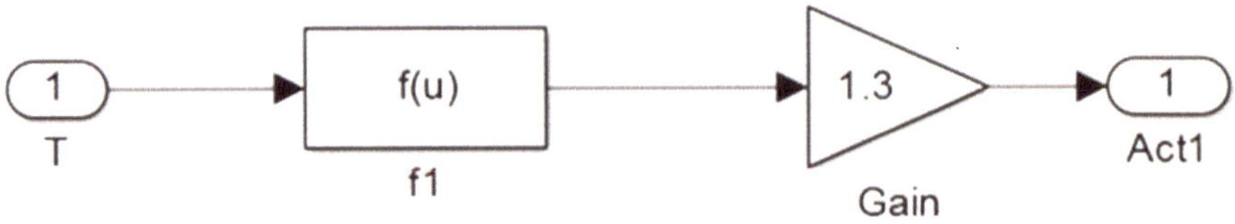

Fig. 5.2 Modelling of activation polarization dependent just on temperature

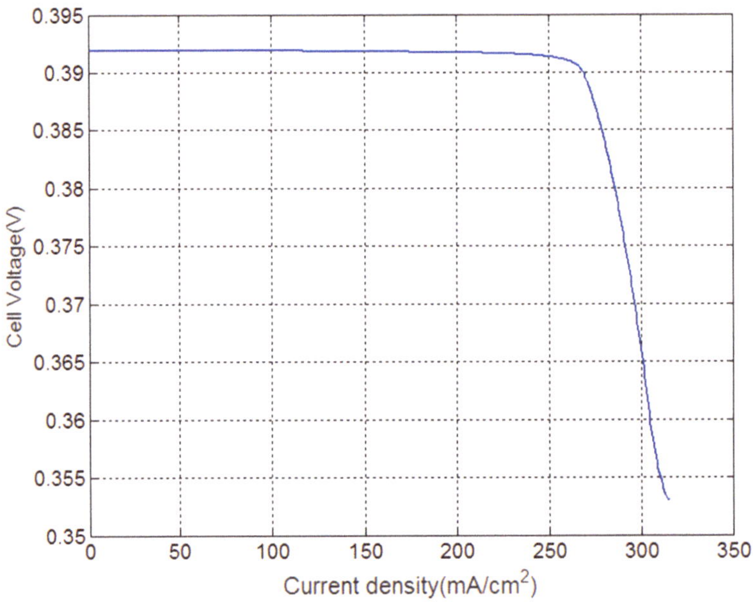

Fig. 5.3 Polarization activation curve dependent only on temperature

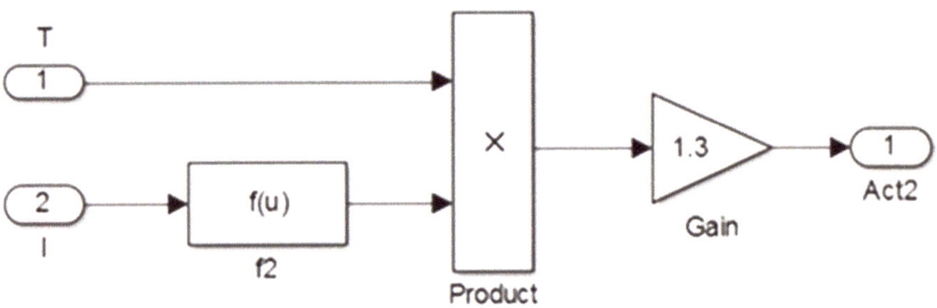

Fig. 5.4 Modelling activation losses dependent on temperature and current density

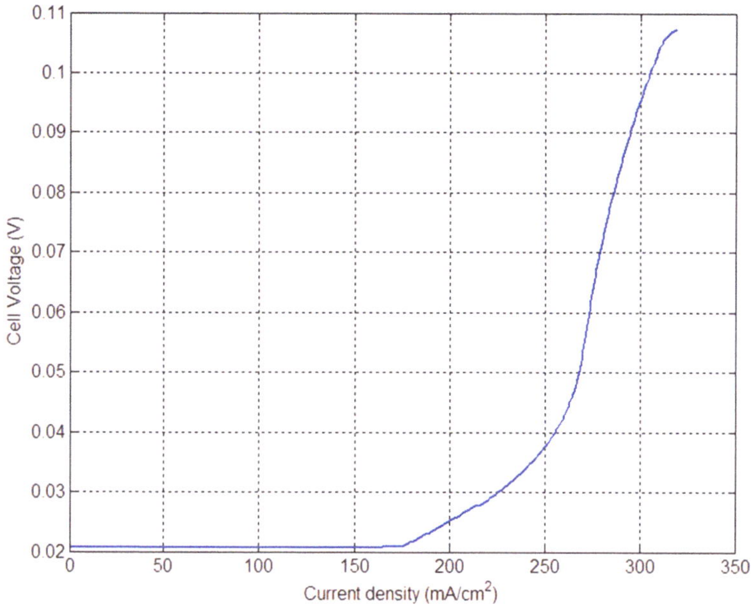

Fig. 5.5 Polarization activation curve dependent on temperature and current density

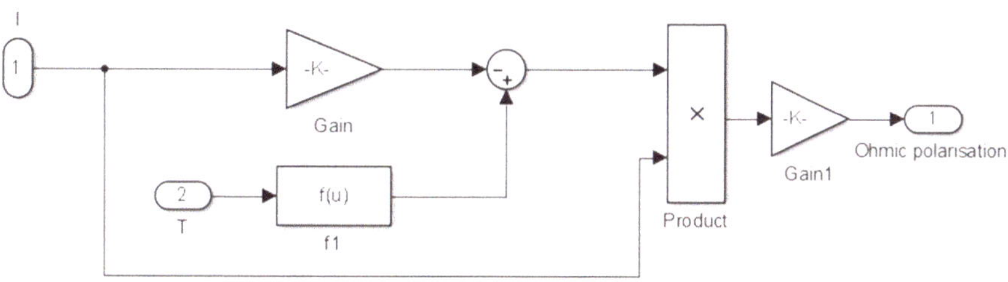

Fig. 5.6 Modelling of ohmic polarization

$$\eta_{\text{ohm}} = iR^{\text{internal}} \tag{5.32}$$

where I represents the current of the cell and

R^{internal} is the internal resistance of the cell and is modelled using the following relationship:

$$R^{\text{internal}} = 0.01605 - 3.5 * 10^{-5} . T + 8.0 * 10^{-5} * i \tag{5.33}$$

5.4.2.3 Modelling of Concentration Polarization

Figure 5.7 illustrates the MatLab Simulink that was used to obtain the concentration polarisation, while Fig. 5.8 illustrates the concentration polarisation of PEM fuel cell at typical operating conditions.

To investigate the influence of temperature on the concentration polarisation, Eq. (3.38) in Chap. 3 is plotted for two values of temperature for the cathode, as depicted in Fig. 5.9. Figure 5.9 shows clearly that the concentration losses increase slowly as temperature increases. This satisfies the Nernst equation. However, this is counterbalanced by the increase in the kinetics of the chemical reaction as predicted

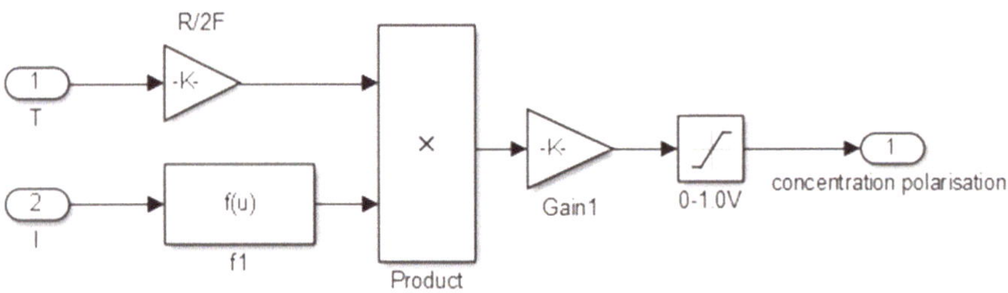

Fig. 5.7 Modelling concentration polarization

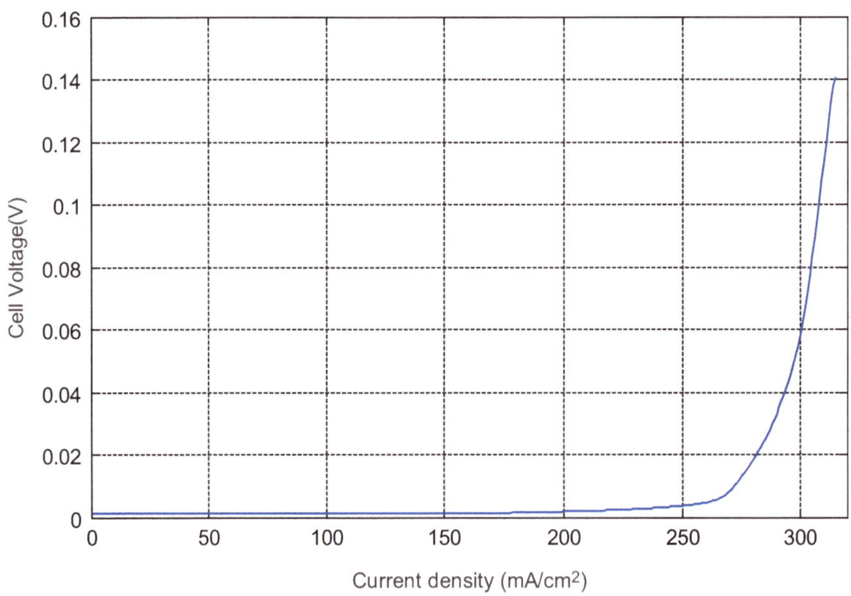

Fig. 5.8 Polarization concentration curve

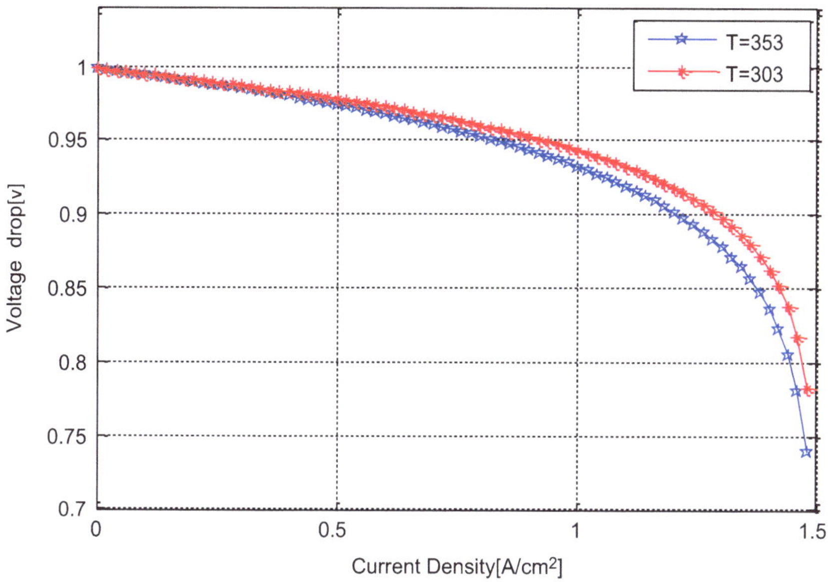

Fig. 5.9 Concentration potential at the cathode side vs. different temperature values

by the kinetic theory of gases. However, the behaviour of concentration losses in FC is a complex phenomenon, and it involves many factors.

The main factor is the water that was produced at the cathode due to the chemical reaction which increases at high current densities; simultaneously, water evaporation increases at higher temperatures, and this causes the reduction of the species transport limitations and thus the concentration losses, but this also could cause the membrane dehydration and reduction in its proton conductivity. Consequently, it is not straightforward to predict the voltage behaviour due to the variation of one parameter without considering the interactions of other parameters, which necessitates the implementation of more advanced analytical tools such as COMSOL modelling and simulations. Such simulation tools can be used to virtually describe the effect of some parameters on the concentration losses and the FC V-I curve. There is a need to consider the correlation among these parameters and their effects on the voltage behaviour.

5.4.2.4 Modelling of Active Pressure

The circuit in Fig. 5.10 is used to simulate the active pressure at both the anode and cathode sides. This model helps to find out the effect of the pressure at the anode and cathode on output voltage of the FC. The pressure of hydrogen at the anode side and oxygen pressure at the cathode side are dependent on current density and temperature.

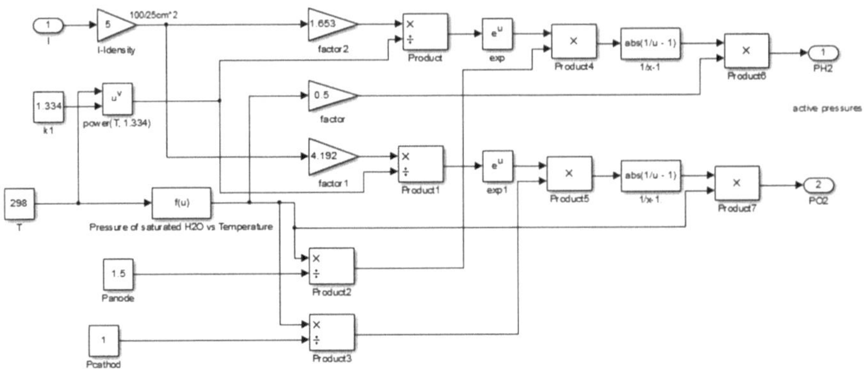

Fig. 5.10 Modelling of active pressure

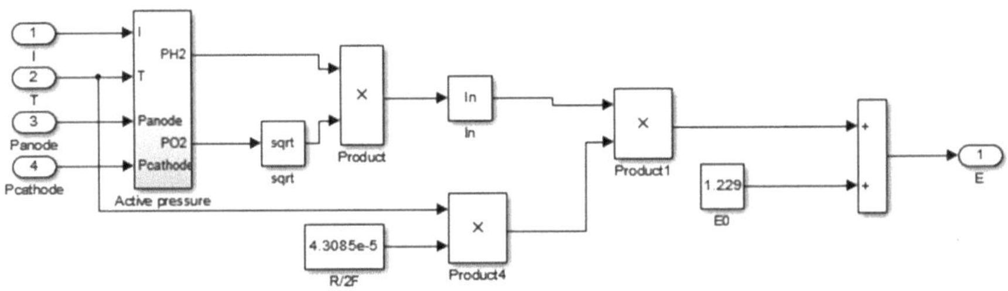

Fig. 5.11 Modelling of Nernst equation

The result of chemical reactions inside a FC is reversible single electrode voltage, E, given by the Nernst equation as shown in Fig. 5.11.

The output voltage of PEM fuel cell depends on parameters that are contained in three major losses and Nernst equation. Equations (5.32) and (5.33) are used to calculate the output voltage of the cell. The final system implementation for the output voltage of PEM fuel cell is tackled in Fig. 5.12.

$$V = E_{\mathrm{o}} - \eta_{\mathrm{act}} - \eta_{\mathrm{ohm}} - \eta_{\mathrm{con}} \tag{5.34}$$

$$\begin{aligned} V = & \left(1.23 - 0.9*10^{-3}(T - 298) + \frac{RT}{4F}\left(\ln\left(\mathrm{P^2H^2}*\mathrm{PO_2}\right)\right)\right. \\ & - \Big[-0.9514 + T*\ 0.00312 + T*7.4*10^{-5}[\ln(C*\mathrm{O_2})] \\ & + 0.000187*\ln(i) + \left(-i*\left(0.01605*\ 10^{-5}\right)* - 3.5*\ T + 8.0*10^{-5}*\ i\right) \\ & \left. - \left(-\beta*\ln\left(1 - \frac{I}{I_{\max}}\right)\right)\right) \Big] \end{aligned} \tag{5.35}$$

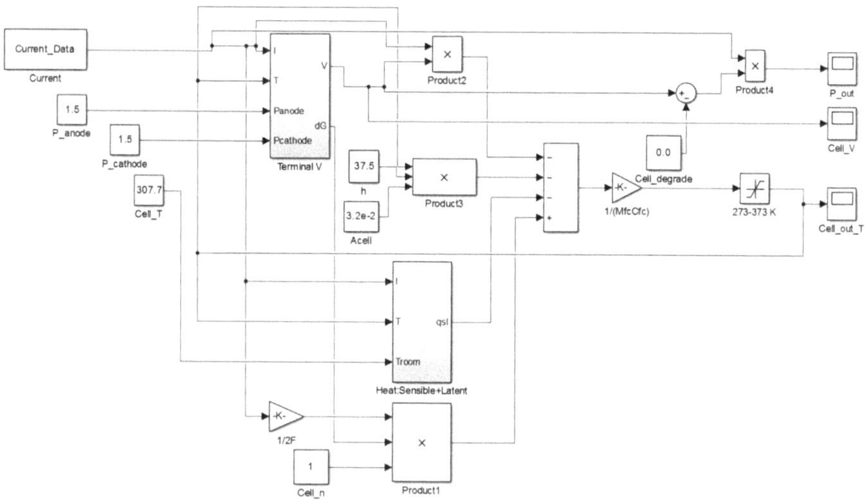

Fig. 5.12 Full model to simulate a proton-exchange membrane fuel cell

5.5 Summary

Modelling is performed to capture the aspects of a PEM fuel cells system that are of interest for study. Knowledge of FC phenomena, such as electrochemical, thermodynamic and transport processes, material properties and various interactions are useful in formulating a model. 3D and 2D models are more accurate in the study of the phenomena that occur in 0D and 1D. However, 3D and 2D need additional time and a computer with high specifications to complete the simulation. A mathematical model, which describes certain aspects of a FC system and predicts its behaviour, may be a set of equations, algebraic or differential, or a computer-based procedure or subroutine.

References

1. Mazumder, S., & Cole, J. V. (2003). Rigorous 3-D mathematical modelling of PEM fuel cells II. Model predictions with liquid water transport. *Journal of the Electrochemical Society, 150* (11), A1510–A1517.
2. Pasaogullari, U., & Wang, C. Y. (2005). Two-phase modelling and flooding prediction of polymer electrolyte fuel cells. *Journal of the Electrochemical Society, 152*(2), A380–A390.
3. Spiegel, C. (2011). *PEM fuel cell modeling and simulation using MATLAB*. Burlington USA: Academic press. ISBN: 978-0-12-374259-9.
4. Alrweq, M., Albarbar, A. (2016). *Investigation into the characteristics of proton exchange membrane fuel cell-based power system IET science, measurement & technology*. doi:https://doi.org/10.1049/iet-smt.2015.0046, Online ISSN 1751–8830.

5. Milewski, J., Świrski, K., Santarelli, M. and Leone, P., 2011. Advanced methods of solid oxide fuel cell modeling. Springer Science & Business Media.
6. Al-Baghdadi, M. A. (2010). *CFD modeling and analysis of different novel designs of air-breathing PEM fuel cells*. New York: Nova Science Publishers.
7. Bavarian, M., Soroush, M., Kevrekidis, I. G., & Benziger, J. B. (2010). Mathematical modelling, steady-state and dynamic behaviour, and control of fuel cells: A review†. *Industrial & Engineering Chemistry Research, 49*(17), 7922–7950.
8. Andersson, M., Yuan, J., & Sundén, B. (2010). Review on modelling development for multiscale chemical reactions coupled transport phenomena in solid oxide fuel cells. *Applied Energy, 87*(5), 1461–1476.
9. Vasile, N. S., Doherty, R., Videla, A. H. M., & Specchia, S. (2016). 3D multi-physics modeling of a gas diffusion electrode for oxygen reduction reaction for electrochemical energy conversion in PEM fuel cells. *Applied Energy, 175*, 435–450.
10. Al-Masri, A., Peksen, M., Blum, L., & Stolten, D. (2014). A 3D CFD model for predicting the temperature distribution in a full scale APU SOFC short stack under transient operating conditions. *Applied Energy, 135*, 539–547.
11. Abdollahzadeh, M., Pascoa, J. C., Ranjbar, A. A., & Esmaili, Q. (2014). Analysis of PEM (polymer electrolyte membrane) fuel cell cathode two-dimensional modeling. *Energy, 68*, 478–494.
12. Siegel, C. (2008). Review of computational heat and mass transfer modelling in polymer-electrolyte-membrane (PEM) fuel cells. *Energy, 33*(9), 1331–1352.
13. Liu, Y., Lehnert, W., Janßen, H., Samsun, R. C., & Stolten, D. (2016). A review of high-temperature polymer electrolyte membrane fuel-cell (HT-PEM FUEL CELL)-based auxiliary power units for diesel-powered road vehicles. *Journal of Power Sources, 311*, 91–102.
14. Hutzenlaub, T., Becker, J., Zengerle, R., & Thiele, S. (2013). Modelling the water distribution within a hydrophilic and hydrophobic 3D reconstructed cathode catalyst layer of a proton exchange membrane fuel cell. *Journal of Power Sources, 227*, 260–266.
15. Carton, J. G., Lawlor, V., Olabi, A. G., Hochenauer, C., & Zauner, G. (2012). Water droplet accumulation and motion in PEM (proton exchange membrane) fuel cell mini-channels. *Energy, 39*(1), 63–73.
16. Wang, X., & Van Nguyen, T. (2010). Modelling the effects of the microporous layer on the net water transport rate across the membrane in a PEM fuel cell. *Journal of the Electrochemical Society, 157*(4), B496–B505.
17. Liu, F., Lu, G., & Wang, C. Y. (2007). Water transport coefficient distribution through the membrane in a polymer electrolyte fuel cell. *Journal of Membrane Science, 287*(1), 126–131.
18. Das, P. K., Li, X., & Liu, Z. S. (2010). Analysis of liquid water transport in cathode catalyst layer of PEM fuel cells. *International Journal of Hydrogen Energy, 35*(6), 2403–2416.
19. Lu, Z., Rath, C., Zhang, G., & Kandlikar, S. G. (2011). Water management studies in PEM fuel cells, part IV: Effects of channel surface wettability, geometry and orientation on the two-phase flow in parallel gas channels. *International Journal of Hydrogen Energy, 36*(16), 9864–9875.
20. Grimm, M., See, E. J., & Kandlikar, S. G. (2012). Modelling gas flow in PEM FUEL CELL channels: Part I–flow pattern transitions and pressure drop in a simulated ex situ channel with uniform water injection through the GDL. *International Journal of Hydrogen Energy, 37*(17), 12489–12503.
21. Gao, F., Blunier, B., Simoes, M. G., & Miraoui, A. (2011). PEM fuel cell stack modelling for real-time emulation in hardware-in-the-loop applications. *IEEE Transactions on Energy Conversion, 26*(1), 184–194.
22. Wang, Y., Chen, K. S., Mishler, J., Cho, S. C., & Adroher, X. C. (2011). A review of polymer electrolyte membrane fuel cells: Technology, applications, and needs on fundamental research. *Applied Energy, 88*(4), 981–1007.
23. Lobato, J., Cañizares, P., Rodrigo, M. A., Pinar, F. J., Mena, E., & Úbeda, D. (2010). Three-dimensional model of a 50 cm 2 high temperature PEM fuel cell. Study of the flow channel geometry influence. *International Journal of Hydrogen Energy, 35*(11), 5510–5520.

Chapter 6
Experimental Set-Up, Results and Data Analysis

6.1 Overview

As explained in Chap. 3, in hydrogen fuel cells (HFCs), the fuel (hydrogen) splits into electrons and protons at the anode. Protons travel through the membrane (electrolyte), while electrons migrate to the cathode via an external circuit. At the cathode, oxygen molecules unite with electrons and protons that have migrated from the anode, resulting in the generation of water and heat. This process comprises and induces a number of chemical and electrical variables which should be experimentally measured and analysed to allow the extraction of useful information. In addition to that, movement of water bubbles and collisions with each other will result in dynamic signatures such as acoustic emissions AE. Good understanding of these parameters will enable better control, monitoring and diagnosis techniques for PEM fuel cells. In this chapter clearer understanding on the effects of operational parameters will be gained through a series of experiments. Results are also validated using the developed finite element model – in COMSOL software package.

6.2 Effect of Temperature and Pressure

6.2.1 Effect of Temperature

Figure 6.1 shows an increase in the temperature that improves the performance of the cell; reduces voltage losses, which leads to a reduction of voltage dropping factors, especially the decline in the activation losses; and improves the overall outcome of the cell. However, under operational conditions, in order to prevent the loss of humidity, which is significantly important to boost the membrane conductivity, the temperature of the polymeric PEM fuel cell does not exceed 80 °C [1].

Figure 6.1b depicted the modelling results of the temperature in the interior layers.

A. Albarbar, M. Alrweq, *Proton Exchange Membrane Fuel Cells*,
https://doi.org/10.1007/978-3-319-70727-3_6

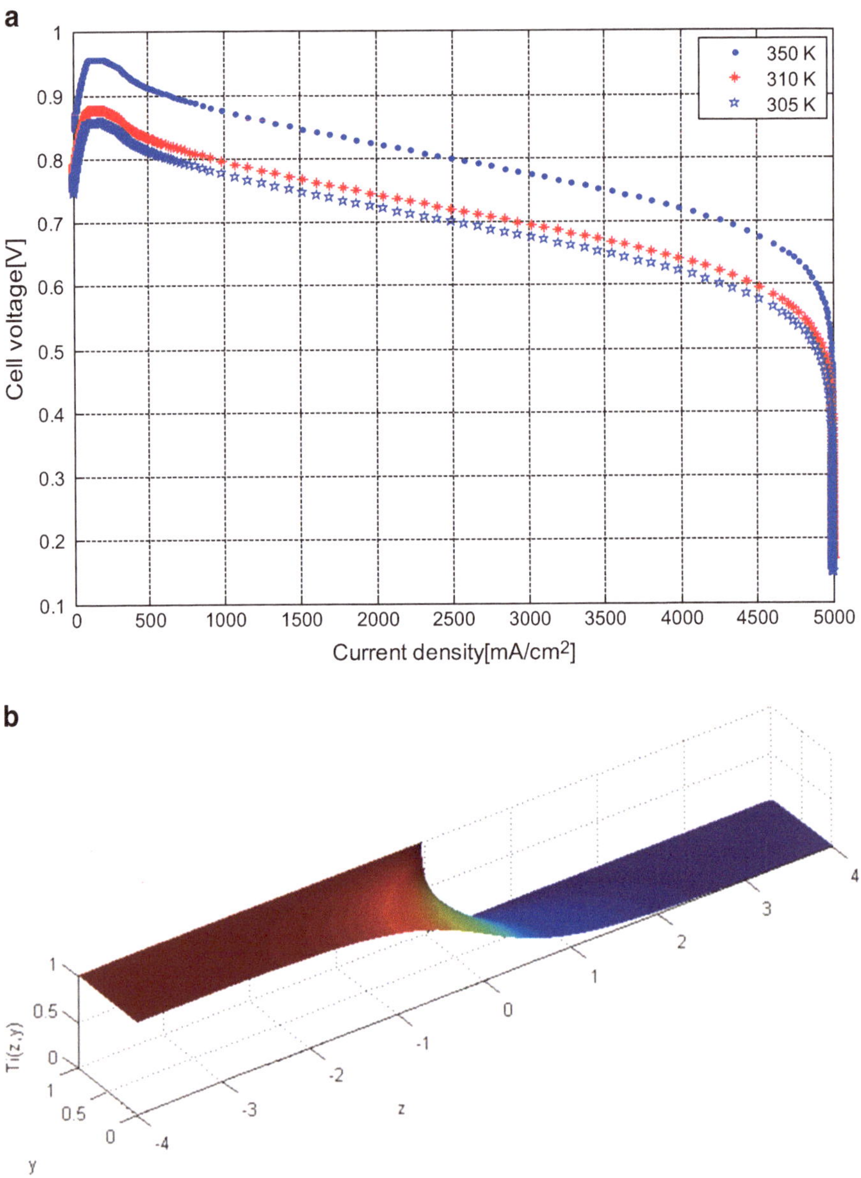

Fig. 6.1 (**a**) The effect of temperature on PEM fuel cell output voltage. (**b**) Temperature in internal layers

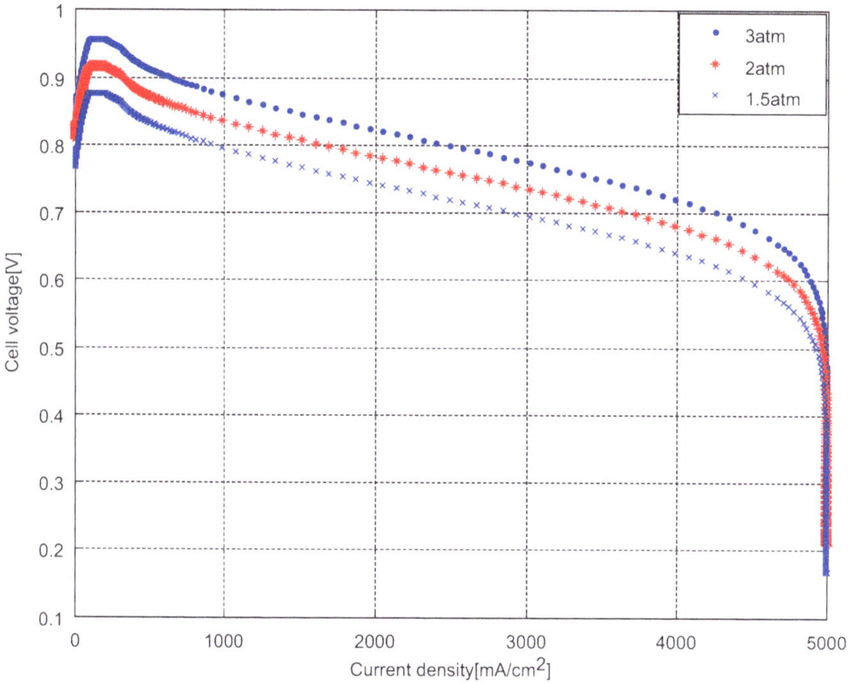

Fig. 6.2 Effect of pressure on PEM fuel cell output voltage

6.2.2 *Effect of Pressure*

Figure 6.2 shows the improvement in the performance of the polymeric PEM fuel cell with three different values of pressure. The increase in pressure values must be within the allowable pressure range as instructed by the data sheet (high pressure value may cause damage to the delicate components, such as the membrane). From the Fig. 6.2, it can be noticed that increased pressure will only improve the performance slightly. This increase in the pressure value decreases temperatures [1].

6.2.3 *Polarisation Curve and Power Density*

Figure 6.3 shows the polarisation curve of the PEM fuel cell for two cases of operation: The first case of works is under the conditions of the cell such as temperature and pressure (353 K &3 atm). The second case of operation is within the statutory requirement (323 k &1.5 atm), which shows the decline in output cell. Polymeric FCs typically operate at a temperature of 70–90 °C and atmospheric

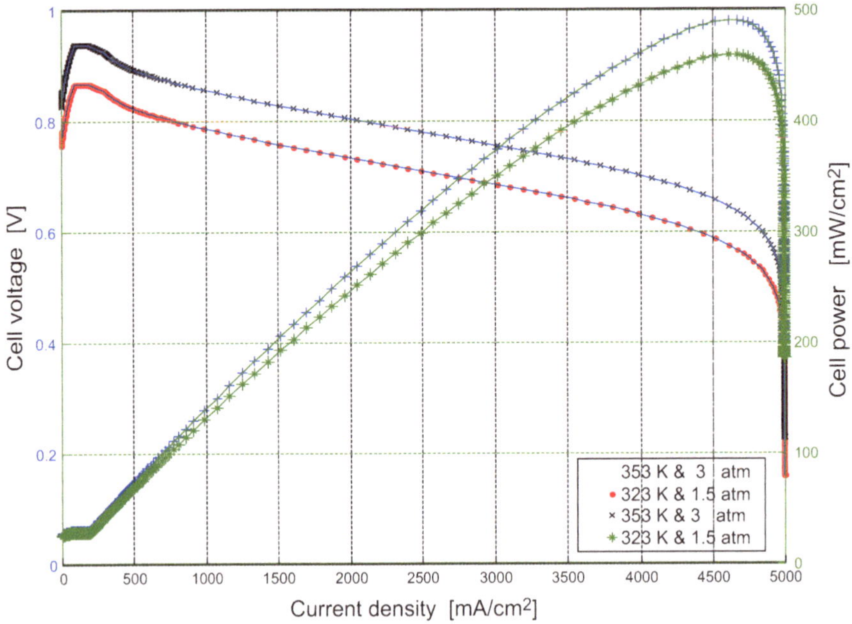

Fig. 6.3 The effect of pressure and temperature on both output voltage and power

pressure (1.5–3.5 atm) since cell gives its highest outcome. Therefore, Fig. 6.3 shows at low temperature and pressure the voltage declines 0.1 V, and it causes the decline in the outcome by 10% of its nominal output.

6.3 Acoustic Emission Phenomena

Water bubbles are generated because of the reaction between hydrogen and oxygen at cathode side. The source of AE signals is the flow and friction between water bubbles [2].

Minnaert resonance equation can be used to find bubble resonance frequencies f_o at particular radius as below:

$$f_o = \frac{1}{2\Pi a}\left(\frac{3\gamma p_a}{\rho}\right)^{1/2} \tag{6.1}$$

while the volume of the bubbles can be determined by the equation below:

$$V = \frac{1}{(6 * \pi 2 * f_{\text{o}})} * \left(\frac{3 * \gamma * p_{\text{a}}}{\rho}\right)^{3/2} \tag{6.2}$$

where a is the radius of the bubble, γ is the polytrophic coefficient, p_{a} is the ambient pressure and ρ is the density of water. Table 6.1 outlines frequencies of the generated acoustic signals associated with bubble sizes. That confirms two important outcomes: the first is the generation of acoustic signals during this process and the second identifies the frequency bands of such generated signals between 100 kHz and 850 kHz.

Figures 6.4 and 6.5 depicted the effect of the different load values on AE density.

Table 6.1 Approximation of the size and volume of bubbles based on frequency information of the average power spectrum

Frequency (kHz)	Bubble radius (m)	Bubble volumes (m3)
100	8.6953e-05	2.7539e-12
200	4.3477e-05	3.4424e-13
300	2.8984e-05	1.0200e-13
400	2.1738e-05	4.3030e-14
500	1.7391e-05	2.2031e-14
600	1.4492e-05	1.2750e-14
650	1.3377e-05	1.0028e-14
700	1.2422e-05	8.0289e-15
750	1.1594e-05	6.5278e-15
800	1.0869e-05	5.3787e-15
850	1.0230e-05	4.4843e-15

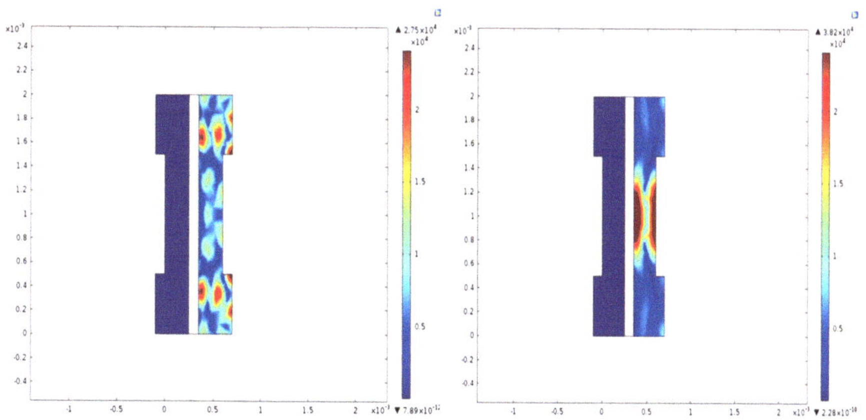

Fig. 6.4 Acoustic emission intensity at (1, 3.3) Ω

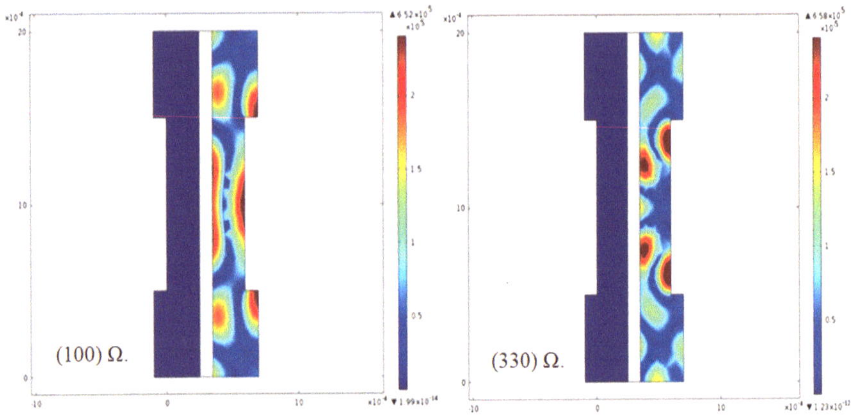

Fig. 6.5 Acoustic emission intensity at (100, 330) Ω

Fig. 6.6 Block diagram of AE signal measuring system

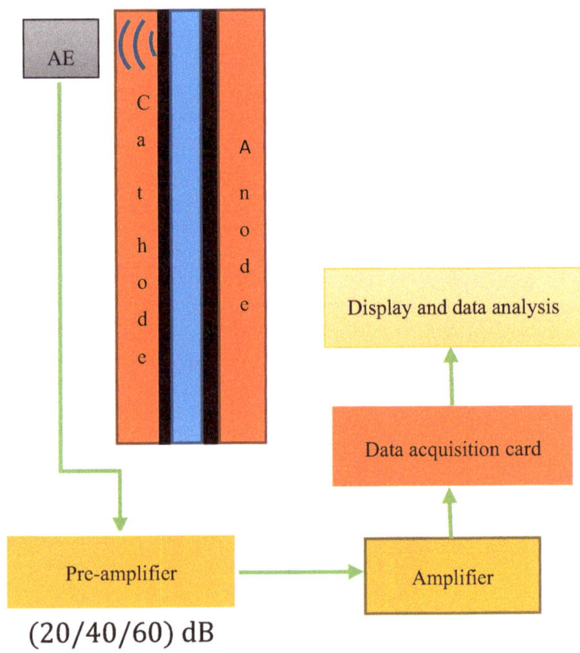

6.3.1 Experimental Set-Up to Investigate the Impact of Load Variation on Acoustic Emission

The experimental set-up is shown in Figs. 6.6 and 6.7. A single 350 mW PEM fuel cell with Nafion 115 membrane of thickness 0.127 mm and active area 9 cm^2 was examined. The electrodes were made of 0.2 mg/cm^2 platinum for both electrodes.

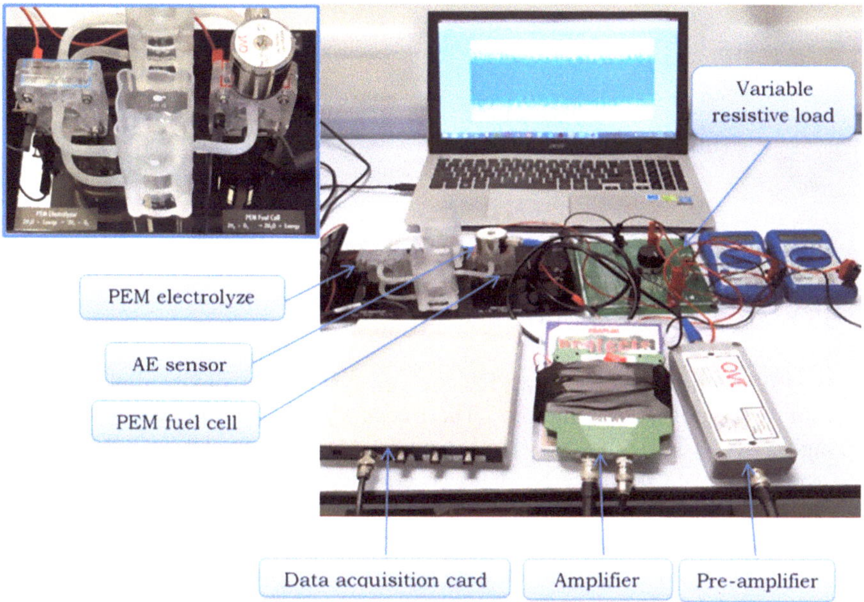

Fig. 6.7 The experimental set-up

The catalyst layer was made of carbon-supported platinum, loading 1 mg/cm^2. The membrane electrode assembly (MEA) was first sandwiched between two gaskets and then between two graphite current collector plates. There were flow channels on each plate surface.

Six different loads were applied in order to investigate the active levels of AE, namely, 20%, 40%, 60%, 80% and 100% of cell's full load.

The AE monitoring was conducted using a piezoelectric sensor mounted atop the PEM fuel cell. The sensor was attached using special wax, which supported the sensor during the experiment while supplying the necessary acoustic coupling.

The AE signals from the PEM fuel cell were measured using an AE transducer model NS2000 and amplified and filtered using 20 kHz to 5 MHz band pass filter. The gain of the amplifier was set to 30 dB and interfaced to a GDS-1102 - GW Instek Digital Oscilloscopes. For all experiments conducted, a sampling interval of 200 ns was used, necessitating 1024 points per signal. The scope was triggered such that about 100 points were collected before the trigger point. The recorded data was processed using MATLAB package [3–5].

6.3.2 AE Measurements and Data Analysis

Statistical approaches were utilised to extracting information from AE data. FFT is used to generate a frequency domain for the representation of the signal. The system

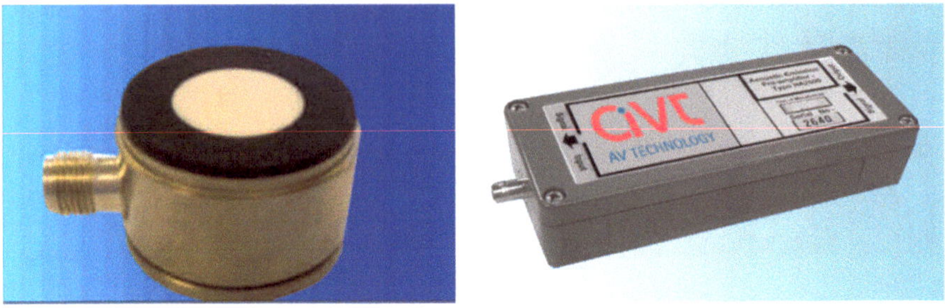

Fig. 6.8 AE sensor and preamplifier

was used to collect data from the PEM fuel cell system that comprises piezoelectric sensor, preamplifier and amplifier. The frequency spectrum of AE is very large (ranging from 16 kHz to 1 MHz), meaning the AE sensor model NS2000m is used to gather the data from the PEM fuel cell system. A preamplifier amplified the piezoelectric sensor with 20–up to 1000 kHz bandwidth filters and a 30 dB gain. See Fig. 6.8 [2].

A fast storage oscilloscope was used as transient digitiser. This can acquire and digitise input signals and display or store the result. GDS-1102 - GW Instek Digital Oscilloscopes is used for monitoring AE and can sample data at rates of up to 1 MHz. Code in MATLAB environment has been written for data analysis in both time and frequency domains [6].

6.3.3 Experimental Procedure of Water Flooding

As mentioned in Chap. 5, water management plays a vital role in enhancing the performance of PEM fuel cell. In this content, single PEM cell utilised to figure out the effect of flooding on cell performance with the following specifications:

- Fuel cell active area: 9 cm^2
- Membrane electrode assembly (MEA) with Nafion 115, 1 mg. Pt/cm^2 (platinum load at both sides)
- Toray carbon fibre paper with 0.19 mm thickness used as GDL

In this experiment, operating temperature, pressure and relative humidity have been taken into consideration due to their significant impact on water management issue.

Hydrogen and oxygen are used as fuel and oxidiser, respectively. Electrolysis is used to ensure both hydrogen and oxygen have a humidity and purity of 100% and 99.9%, respectively. There are two approaches to making the cell operate under flooding conditions. The first method relies on the difference between the temperature of the gases and the cell, i.e. if the cell's temperature is higher than the temperature of gases, the cell will remain under dry condition. In contrast, if the gas's temperatures are higher than the cell's temperature, the cell will be flooded. The second method depend on control the water drainage. In the existing work, the authors opted for the second method due to its simplicity.

Fuel cell cleaning from accumulated water can be performed in two ways, one is by raising the level of the cell temperature and the second method uses the air stream. Both these methods have some limitations. For example, raising cell temperatures tend to limit cell's applications, especially mobile applications. The second method that relies on air stream may damage the delicate components such as membrane. Thus, this method should be used carefully and prudently. In this work, the second method was used. To ensure the repeatability, several independent flooding experiments were carried out.

In order to characterise the flooding condition, and simplify the experiments' procedure, we assume that the flow channel of the PEM fuel cell is signal-phase flow, so that the flooding condition can be declared if the measured values of current and voltage drop deviate significantly from its predicted values towing to the presence of liquid water in the system.

The output PEM fuel cell connected to different electrical load in order to vary the current drawn from the system.

The data that is obtained from output voltage and current is then fed into a personal computer running a developed computer program under LabVIEW software environment to monitor the water content and cell flooding in every moment.

6.4 Results and Discussion

6.4.1 Effects of Operating Conditions on the AE Signals

Firstly, the electrical performance of the PEM fuel cell was examined by connecting it to different loads (resistors). The effects of electric loading variations on the generated currents, voltages and power are listed in Table 6.2, which was as expected and demonstrates a healthy performance.

The output of an AE transducer attached on the 350 mW PEM fuel cell with no load is shown in Fig. 6.9. The AE signals shown in Figs. 6.9 and 6.10 were enhanced by averaging. In practice, it is frequently the case that with a repeated signal, the signal to noise ratio can be improved by averaging, particularly where the corruption of the signal is due to unwanted noise occurring because of random events. Time-domain averaging is a way to reduce the content of undesired components in a signal.

The main feature could be observed from the AE waveform, and shown in Fig. 6.10 is peaks corresponding to the reaction event between oxygen and hydrogen at the cathode. This waveform is complicated and difficult to extract information from due to the numerous frequency components superimposed on each other. In the associated power spectrum, shown in Fig. 6.10, there cannot be seen any peaks at 650 kHz; the other peaks that appear are results of random events.

Table 6.2 Load effects on power parameters

Resistance (Ohr	Current (A)	Voltage (V)	Power (W)
∞	0	0.99	0
330	0.01	0.97	0.0097
100	0.02	0.94	0.0188
33	0.03	0.9	0.027
10	0.08	0.83	0.0664
3.3	0.22	0.77	0.1694
1	0.56	0.63	0.3528
0.33	1.05	0.47	0.4935
0.1	1.43	0.32	0.4576
0	1.61	0.24	0.3864

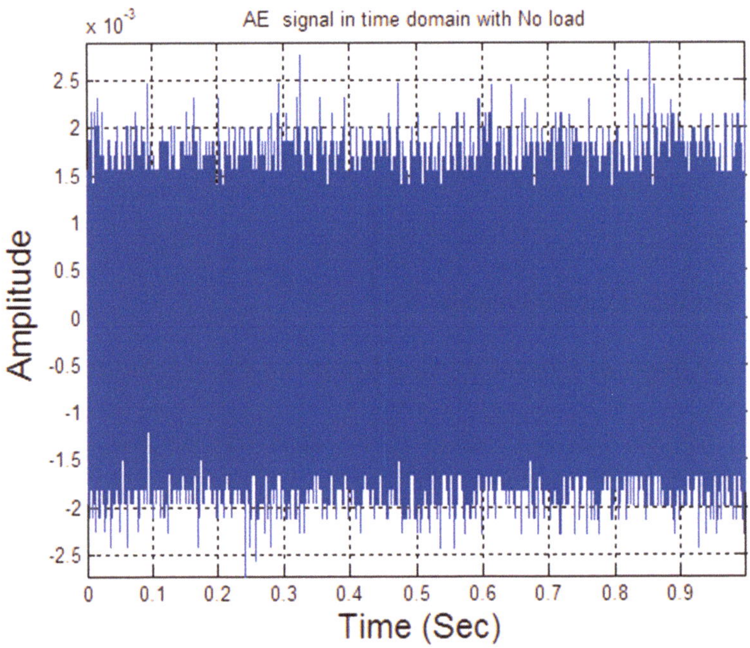

Fig. 6.9 AE signals in time domain with no load

This section reports on how varying the PEM fuel cell electric load affects the characteristics of the AE signals. The cell was operated under six loading condition. A MATLAB code was written to show the measured AE signal on terms of both the time domain and frequency domain [2]. Extensive experiments have been made to validate the AE activity with load change. The effect of loads changing on AE activity is described in Figs. 6.11, 6.12 and 6.13. The excellent correlation between load changing and AE has been observed.

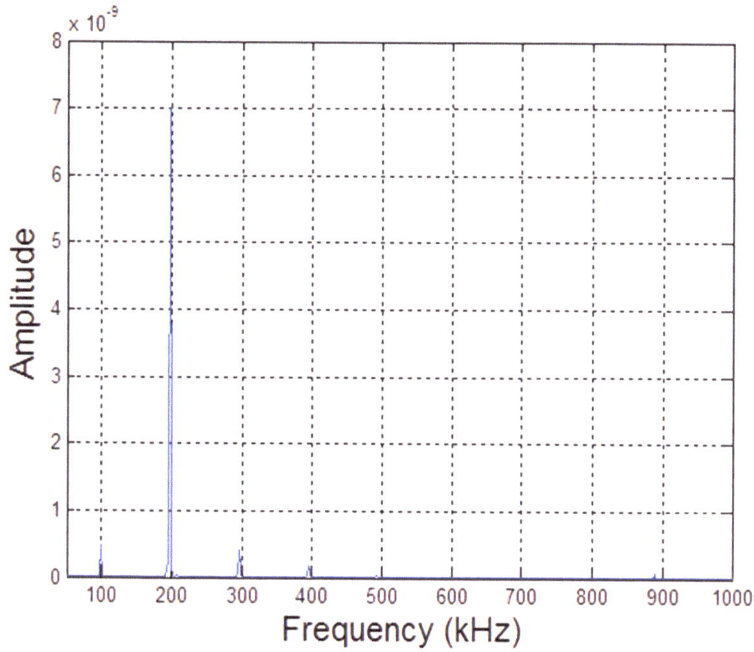

Fig. 6.10 AE signals in frequency domain with no load

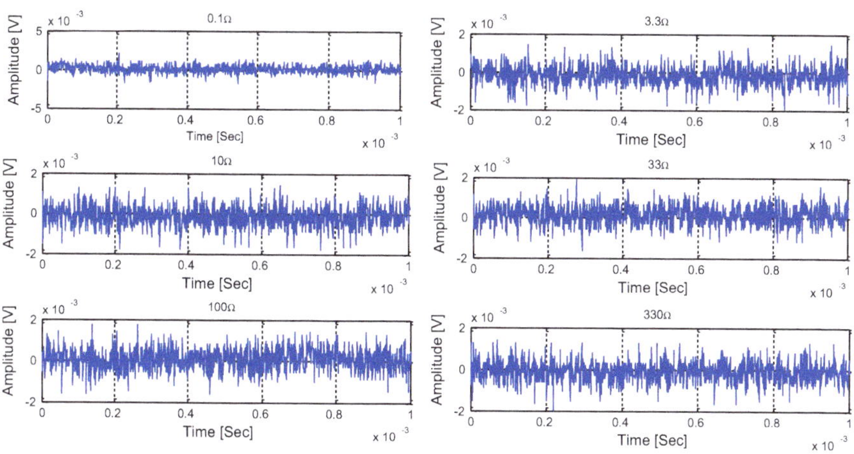

Fig. 6.11 Shows time-domain histories at each of the six loads with the associated spectrum

Figure 6.11 shows time-domain histories at each of the six loads with the associated spectrum. It can be observed that the amplitudes of the time domain waveforms increase as the load increases. Figure 6.13 shows the AE signals in the

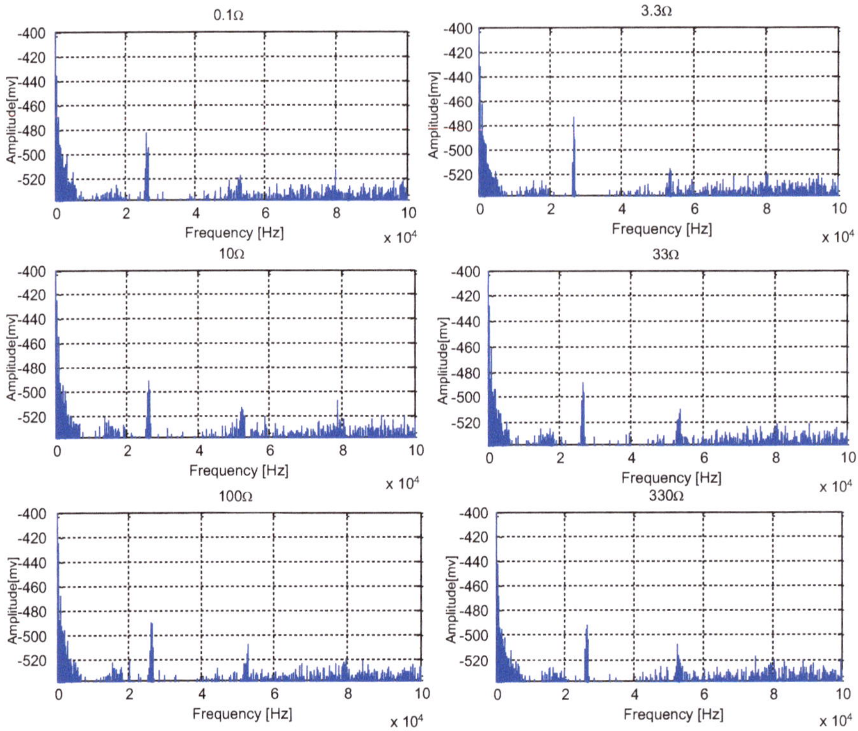

Fig. 6.12 Shows frequency-domain histories at each of the six loads with the associated spectrum

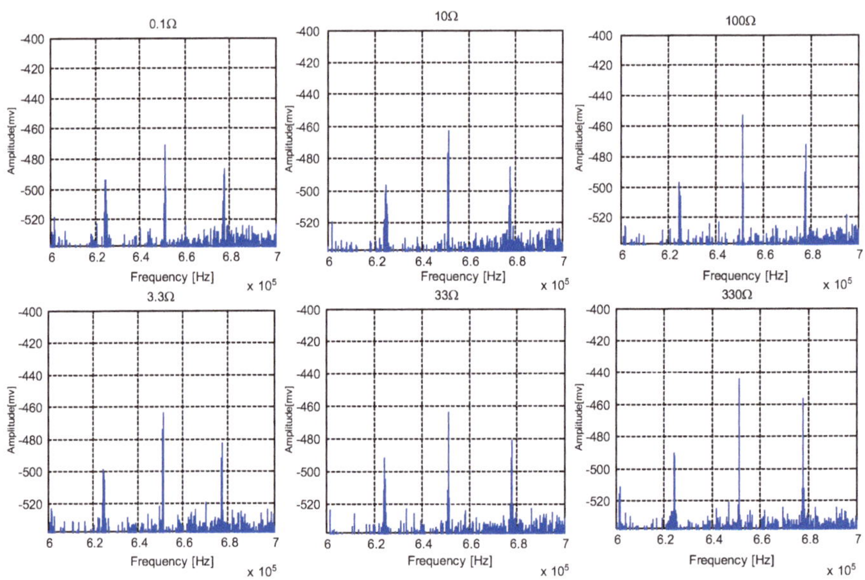

Fig. 6.13 AE signals in frequency domain after the zoom

frequency domain. In the associated power spectrum, shown in Fig. 6.12, the peaks can be seen, the highest at the resonance frequency of the AE transducer (650 kHz) and the other peaks are multiples of this main frequency component. The amplitudes of any higher harmonics can be ignored because they contain considerably less energy than the first leading terms. The amplitude of the peaks increase with increasing load, and the frequency at which the peak occurs also increases in direct proportion to the increase in load. Figure 6.13 illustrates the AE signal in frequency band after the zoom. It is clear that the peak of the signal is at 650 KHz.

6.4.1.1 AE Signals Analysis Using Statistical Parameters

In this section, statistical parameters were used to extracting information from the AE signal and used to determine the PEM fuel cell with load variation.

The idea behind testing the PEM fuel cell in this way is to investigate the possibility of utilising AE measurements and simple statistical methods for use as a condition-monitoring tool. Calculating the RMS values, kurtosis and skewness of the AE signals may give a quick indication of the PEM fuel cell in a relatively straightforward way, and information on cell's condition could be assessed without the need for any special training [7].

The variance of a population (σ^2) is defined as the sum of squares of the deviations of the observations from the mean divided by the total number of observations; see Eq. (6.3). The standard deviation (σ) is defined as the positive square root of the variance when the standard deviation is evaluated from a sample, rather than the population as a whole; it is given the symbol, s, and (N-1) rather than N that is used in Eq. (6.3).

$$\sigma^2 = \frac{\sum f_i (x_i - \mu)^2}{N} \tag{6.3}$$

Kurtosis and skewness are statistical parameters that could be used to describe the graphical representation of the AE signal population. Kurtosis describes the rate of change of the curve, and skewness reflects the degree to which the curve is skewed to the left or right.

Kurtosis (k) characterises the relative peak ends or flatness of a distribution compared to the Gaussian distribution. Positive kurtosis indicates a relatively peaked distribution. Negative kurtosis indicates a relatively flat distribution.

Skewness characterises the degree of asymmetry of the distribution around its mean. Positive skewness indicates a distribution with an asymmetric tail extending towards values that are more positive. Negative skewness indicates a distribution with an asymmetric tail extending towards values that are more negative. Normal distributions produce a skewness statistic of zero.

The normal distribution is represented by:

$$f(x) = \frac{1}{\sqrt{2\pi\sigma^2}} e^{-(x-\mu)^2/2\sigma^2} \tag{6.4}$$

The formula for evaluating kurtosis (k) is:

$$k = \frac{\Sigma_{i=1}^{N} \left(x_i - \bar{x}\right)^4 / N}{\left(s^2\right)^2} \tag{6.5}$$

where $\Sigma_{i=1}^{N}$ has its usual meaning of the summation of terms from 1 to N, $\bar{x}$ is the sample mean, x_i is the value of the ith score, N is the total number of scores in population and s^2 is the sample variance.

The formula for measuring skewness (sk) is:

$$\text{sk} = \frac{3\left(\bar{x} - \text{md}\right)}{s} \tag{6.6}$$

where *md* is the median score and s is the sample standard deviation.

The root mean square (RMS) of a signal is given by:

$$V_{\text{RMS}} = \sqrt{\frac{1}{T} \int_{t_0}^{t_0+T} v(t)^2 dt} \tag{6.7}$$

where $v(t)$ is the instantaneous value of the signal at time t and V_{RMS} is the RMS value of $v(t)$ for the time T.

6.4.1.2 Root Mean Square Value and Variance as Fault Severity Indicators

In this section, RMS values, kurtosis, skewness and variance are used to investigate the PEM fuel cell's loading effects on their emitted acoustic signals. The PEM fuel cell was tested under six loading conditions, one with no load and four with linear loading increase introduced. The AE data was band pass filtered into three different frequency bands, low-frequency band [0–100 kHz], medium-frequency band [100–500 kHz] and high-frequency band [500–1000 kHz] (Figs. 6.14 and 6.15).

6.4.2 Results and Comments for Water Flooding

Figure 6.16 shows the flooding of PEM fuel cell at constant pressure, fully humidified gases and a slow increase in the current. At current low density, there was no

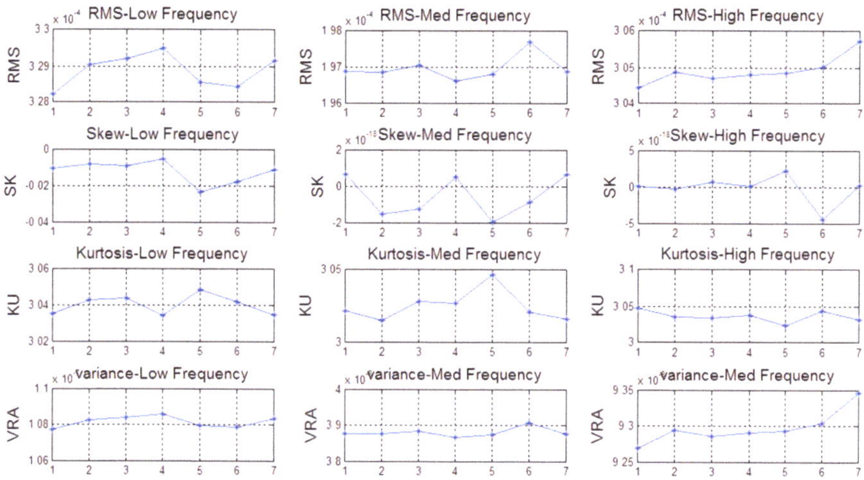

Fig. 6.14 Shows statistical parameters of the AE under different loading conditions

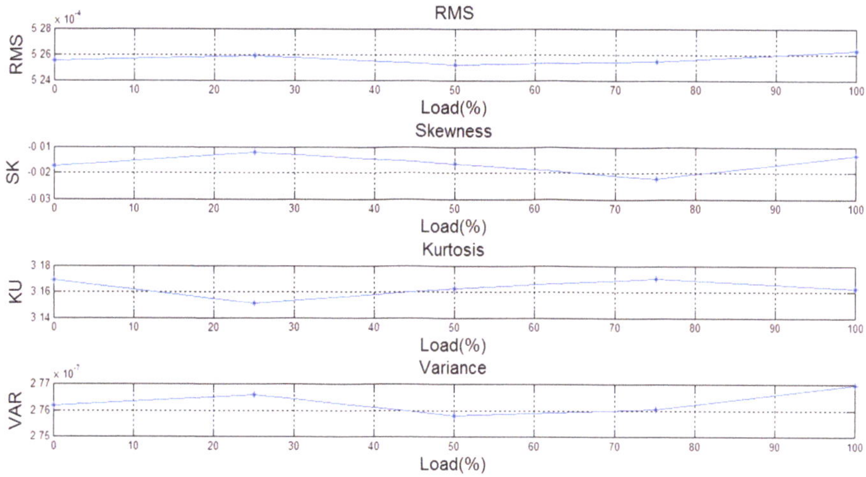

Fig. 6.15 Shows statistical parameters of the AE under different loading conditions at high frequencies

liquid water in the channels, and only a small amount of water was generated at the cathode side due to a reduction in oxygen reduction reaction (ORR). The output voltage and current curves were the same as in normal operating conditions. At 180 s, the first water droplet emerged, which was stressed by three forces: gravitation (G), viscous force between droplet and solid surface (Fs) and shear drag force induced by the gas flow (FD). The droplet was small at the beginning, FD +G$\leq$Fs,

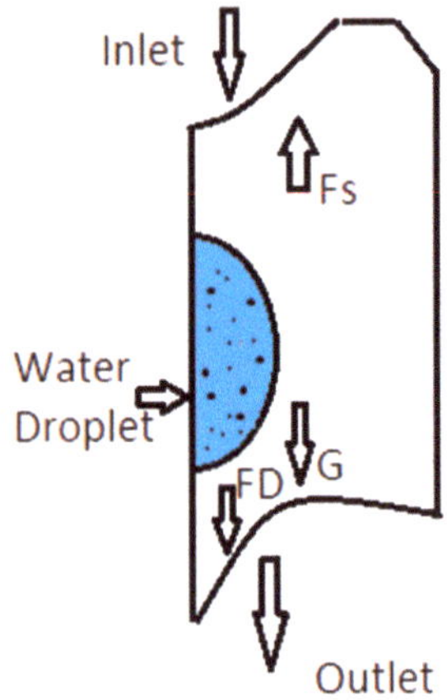

Fig. 6.16 Force diagram of water droplet in the channel

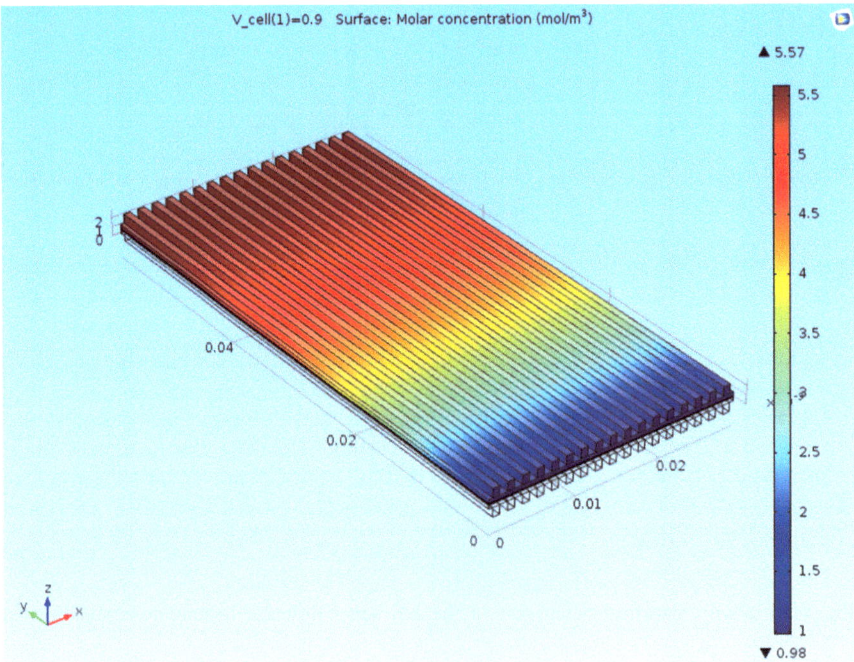

Fig. 6.17 Water concentration at the cathode side

and was held on the GDL surface or/and the channel sidewall. With continued operation of the PEM fuel cell, more droplets emerged and became larger, which started to block the porous membrane and GDL, thereby hindering gas flow.

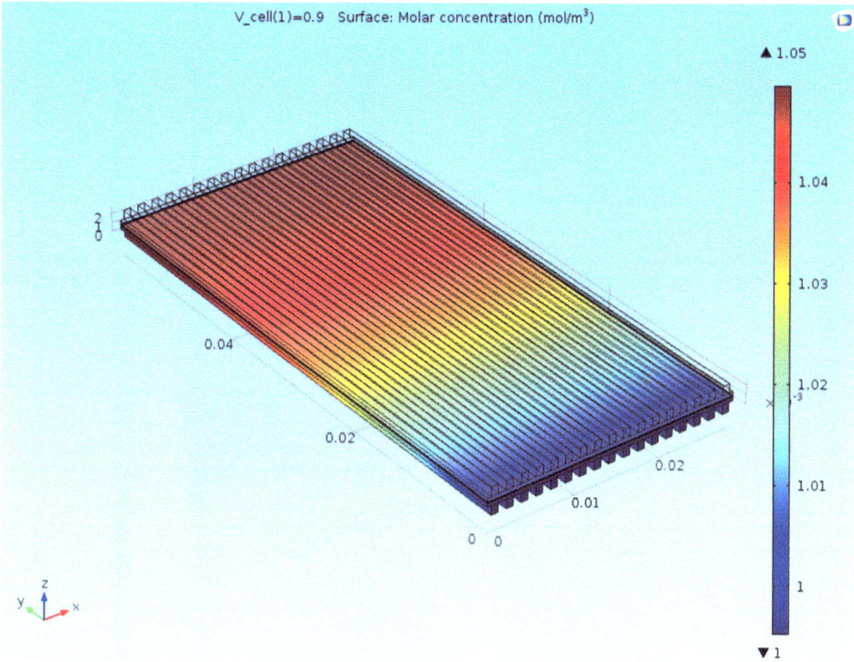

Fig. 6.18 Water concentration at the anode side

Figure 6.17 illustrates the water content at cathode side. From Fig. 6.17, it can be noticed that the level of water at the outlet is higher than the inlet, due to pressures at the outlet will be lower than at the inlet.

Figure 6.18 shows the presence and accumulation of liquid water in anode side. Due to the lack of water generation at the anode, the accumulation of water is attributed to the gas humidifying and back-diffusion phenomenon; hence, the amount of water is less than that at cathode.

The membrane entails more PEM fuel cell components affected by water management issues. The membrane requires adequate water to insure high ionic conductivity, thus lowering ohmic losses.

Figure 6.19 shows the water content at the membrane. The water accumulates in the membrane due to electro-osmotic drag as well as the back-diffusion phenomenon.

Water vapour concentration is depicted in Fig. 6.20. This figure shows that most of the water in liquid phase attributed to PEM fuel cell operates at low relative temperature.

Figures 6.21, 6.22 and 6.23 show water concentration values that are obtained from MATLAB Simulink at the anode, cathode and membrane.

Figure 6.24 depicts the pressures of hydrogen, oxygen and water during cell in flooding condition. It is clear that the pressure of water is increased with flooding

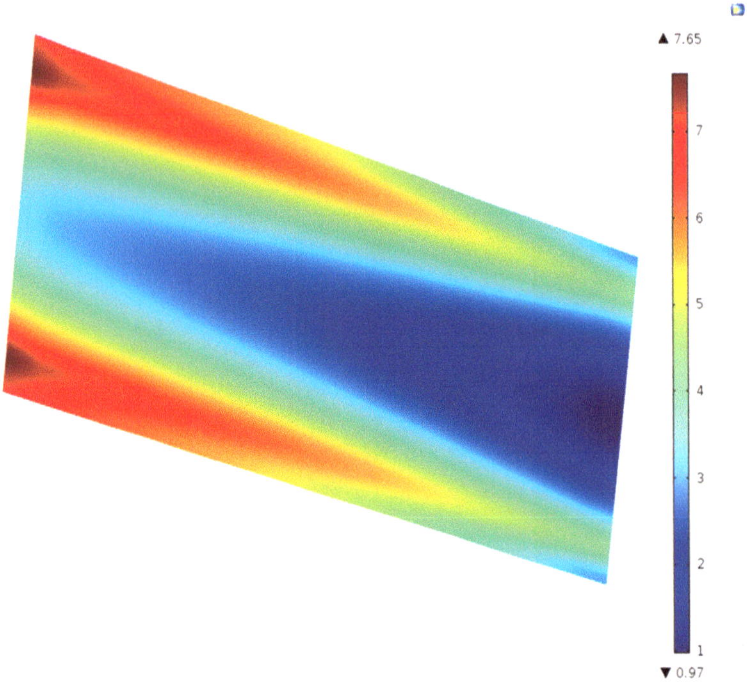

Fig. 6.19 Water content through the membrane

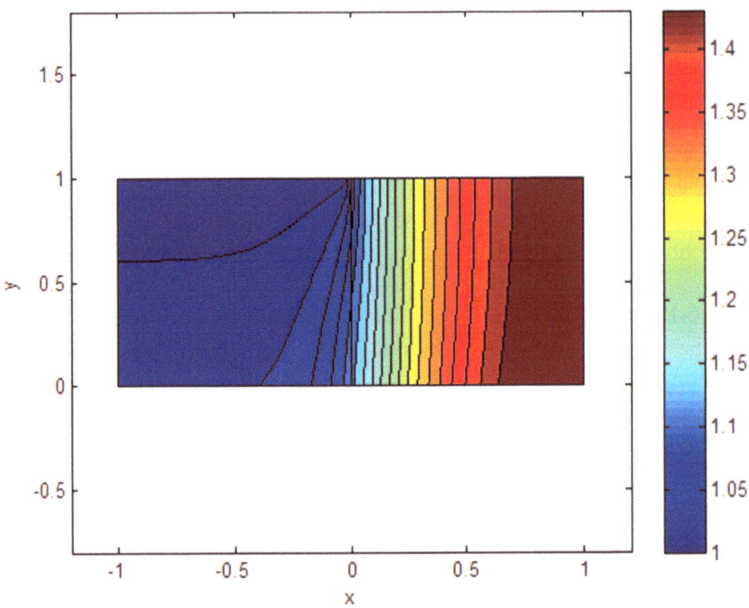

Fig. 6.20 Contour plot of water vapour concentration in membrane

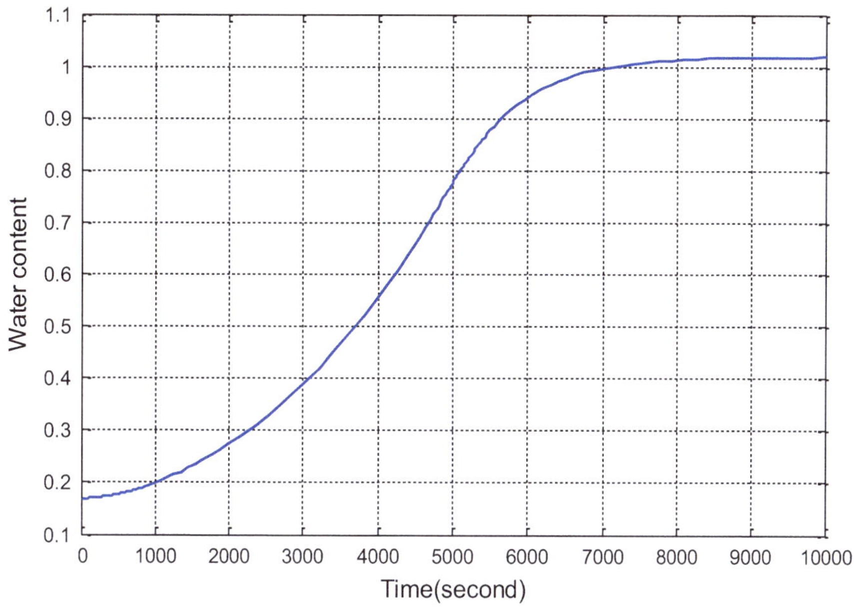

Fig. 6.21 Water concentration at anode

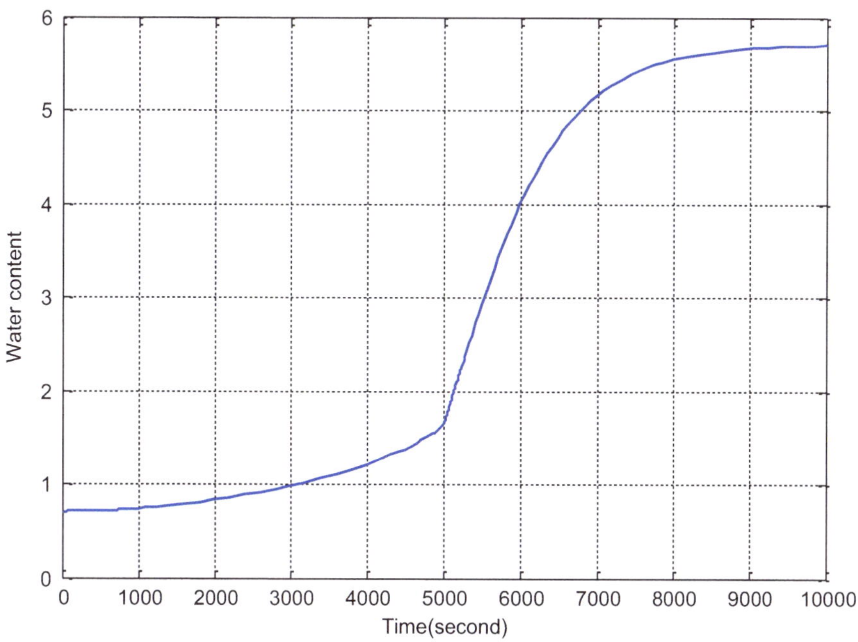

Fig. 6.22 Water concentration at cathode

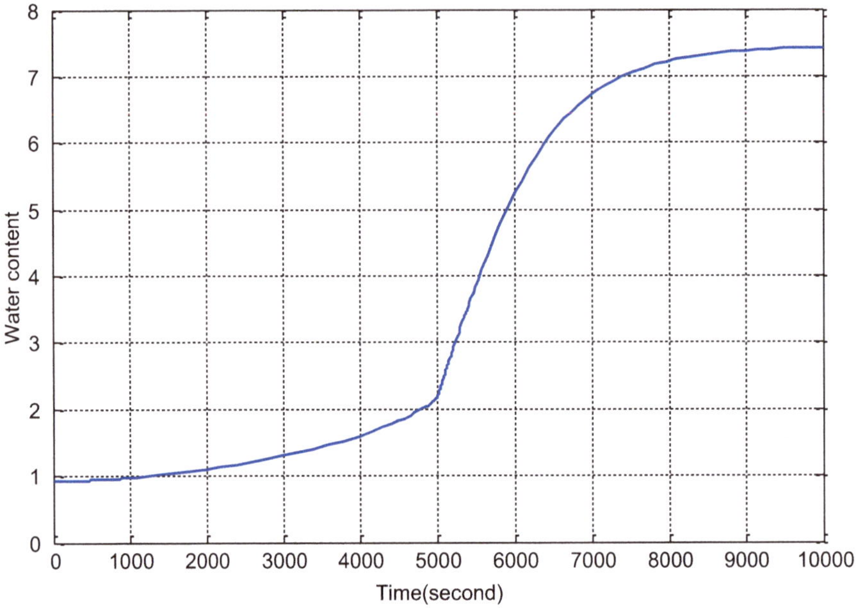

Fig. 6.23 Water concentration at membrane

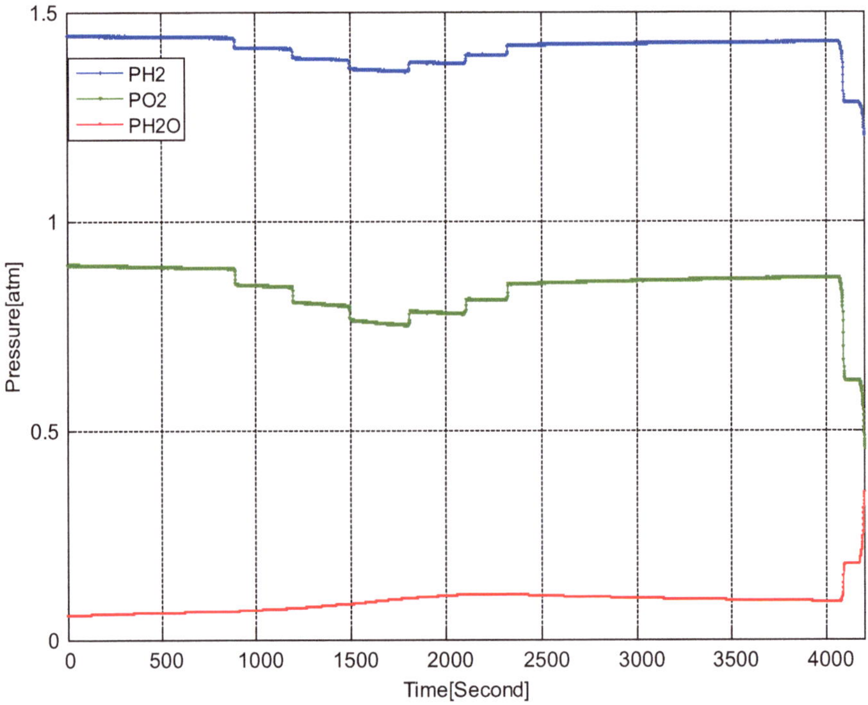

Fig. 6.24 Pressures during flooding

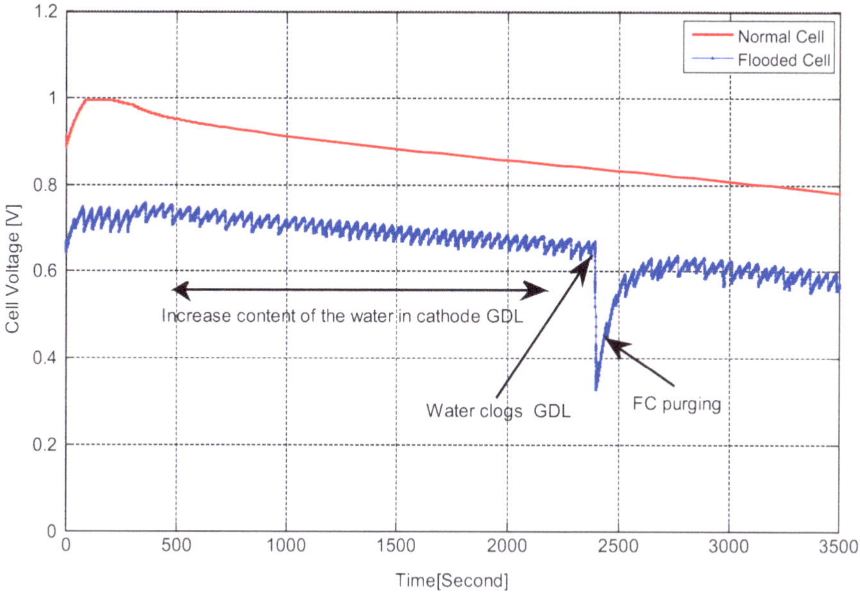

Fig. 6.25 Fuel cell output voltage during flooding state

while reducing the reassurance of hydrogen and oxygen . This is because the water droplets will clog the GDL and gas channels.

Figure 6.25 shows the output voltage of PEM fuel cell under normal conditions as well as during flooding cases. The voltage curve in flooding condition can be divided into three parts: the first part describes the water produced and accumulated at cathode side (cathode side is the main source of the water because of the electrochemical reaction as described in Chap. 1 Eq. 1.1c). As it can be observed from the Fig. 6.25, water takes time to accumulate at the cathode side. This duration depends on many factors, such as current density, pressures and temperatures. The second part presents water status in GDL. When the water clogs GDL, the voltage value declines sharply. Meanwhile, the last part in the curve illustrates the voltage recovery after water purging.

As mentioned in Chap. 3, when the protons are traveling from anode to cathode, they carry water molecules with them, which is known as the electro-osmotic drag phenomenon. The amount of water molecules transferred depends significantly on the value of load. Figure 6.26 illustrates output voltage when different loads are applied on the cell.

Figure 6.27 shows the cell output voltage curves that are obtained from both simulation and experimental results. From Fig. 6.27 it can be noticed that there are reasonable correlation between the experiment and simulation results.

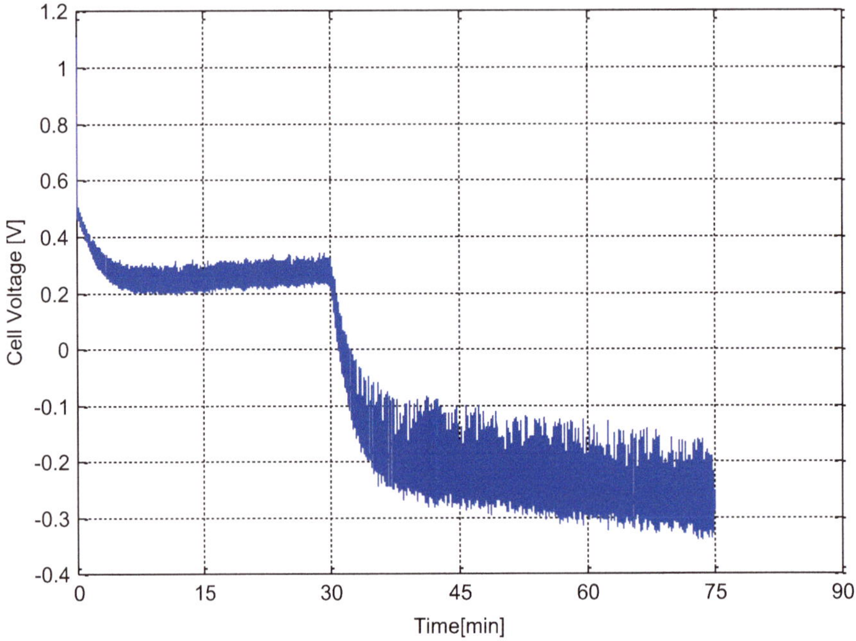

Fig. 6.26 Cell flooding due to load

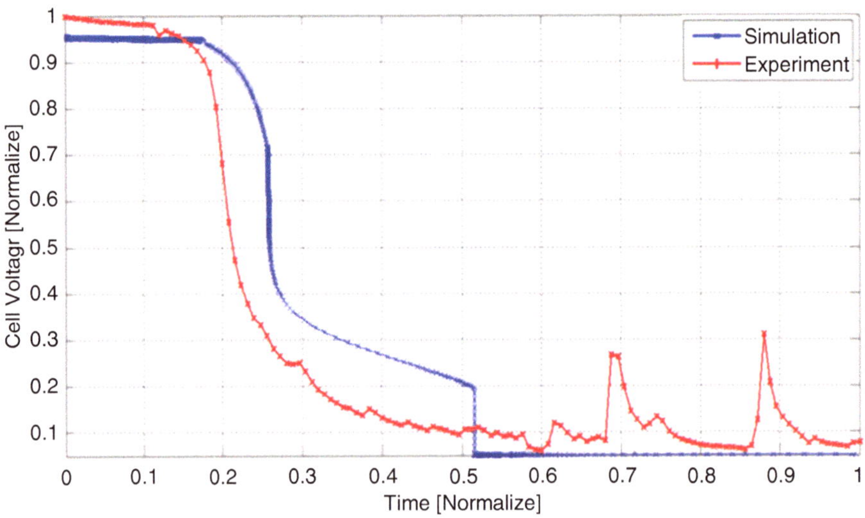

Fig. 6.27 Validation of experiment results

References

1. Alrweq, M., Albarbar, A. (2016). *Investigation into the characteristics of proton exchange membrane fuel cell-based power system IET science, measurement & technology*. doi: https://doi.org/10.1049/iet-smt.2015.0046, Online ISSN 1751–8830.
2. Albarbar, A., Alrweq, M. (2016). Proton exchange membrane fuel cells: Recent developments, modelling and challenges. *Proceeding of 6th International Symposium in Energy Challenges and Mechanics, Inverness.*
3. Albarbar, F. G., Ball, A., & Starr, A. (2010). Acoustic monitoring of engine fuel injection based on adaptive filtering techniques. *Applied Acoustics, 71*(17), 1132–1141.
4. Albarbar, A. (2013). Diesel engine air-borne acoustic signals analysis using continuous wavelet transform. *Advances in Applied Acoustics, 2*(3), 77–82.
5. Albarbar. (2013). An investigation into diesel engine air-borne acoustics using continuous wavelet transform. *Mechanical Science and Technology, 27*(9), 2599–2604.
6. Albarbar, A., Gu, F., Ball, A., & Starr, A. (2010). Diesel engine fuel injection monitoring using acoustic , 2010. *Journal of Measurement, 43*(10), 1376–1386.
7. Teay, S., Batunlu, C., & Albarbar, A. (2017). Smart sensing system for enhancing the reliability of power electronic devices used in wind turbines. *International Journal on Smart Sensing and Intelligent Systems, 37*(10), 407–424.

Chapter 7
Guide to Modelling and Simulation

7.1 Selection of Model

The procedures to develop realistic model for PEM fuel cells using COMSOL software is explained [1]. It starts by selecting suitable three-dimensional interactions between mass transport and electrochemical kinetics [2, 3]. Open COMSOL, choose the model wizard and then click on 3D icon as shown in Fig. 7.1.

Select mass and species transportation as well as the electrochemical kinetics.

Secondary Current Distribution
Secondary current distribution is used to study the electrode kinetics. The reaction kinetics are expressed by Butler–Volmer equation where the mass transfer effects are neglected:

$$i = i_0 \left[\mathrm{e}^{\frac{\alpha n \eta_{\mathrm{actc}}}{RT}} - \mathrm{e}^{\frac{1(1-\alpha) nF \eta_{\mathrm{acta}}}{RT}} \right]$$

To set up the secondary current distribution:

1. Choose the electrochemistry from physics tree.
2. Click on electrochemistry icon to open the list of the products, and then select secondary current distribution (see Fig. 7.2).

7.2 Reactions and Flow in Porous Media

In a PEM fuel cell, the reactants are H_2 on the anode and O_2 on the cathode, and the reaction product is water. The inert phase is N_2; this gas is present in air and hence in the cathode part of fuel cells which operates in air. They never participate in the reactions, but they do affect the transport of mass inside the body of the fuel cell. Mass transport in the anode is excluded from the models used, as the phenomenon

A. Albarbar, M. Alrweq, *Proton Exchange Membrane Fuel Cells*,
https://doi.org/10.1007/978-3-319-70727-3_7

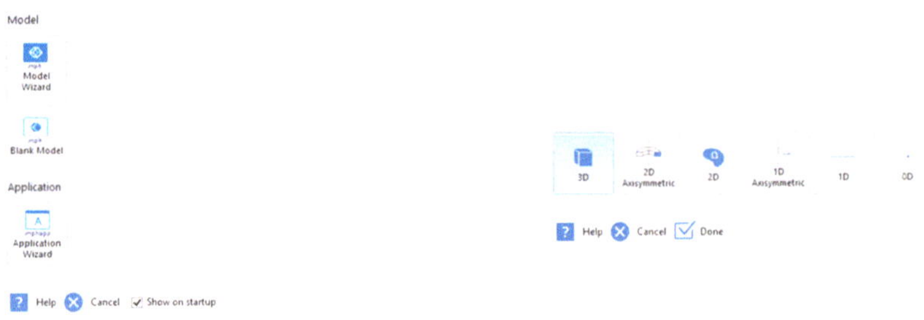

Fig 7.1 Selection of 3D model

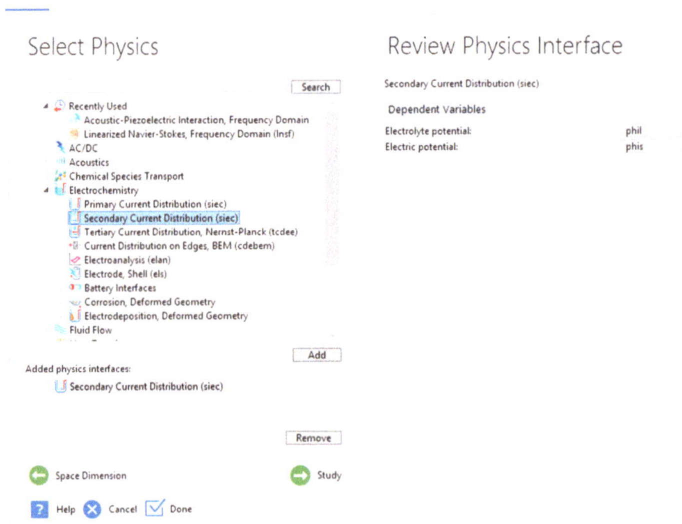

Fig 7.2 Selection of secondary current distribution

of performance limiting usually occurs on the cathode side because of the slower type of reaction kinetics mechanism.

Species are transported on to the cathode through convection as well as the diffusion method. A pressure difference of an external nature is applied across the fuel cell to create the flow. In the free convection case, density variations are produced in the air, caused by differences in concentration and temperature. The reactants are consumed on the electrode, and reaction products are generated on the electrode as well, which creates a density gradient and concentration gradients. The fluid flow passes into the channels, the electrode and the diffusion layer of porous gas. The membrane can transport water, but it is impermeable to gases.

7.2.1 *Reactions and Flow in Porous Media (Anode)*

From physics tree click on reacting flow in porous media, and then select transport of concentrated species (rfcs). The species that transport in anode side are hydrogen and water. So, in the table, write (w_H2 and w_H2O) for the mass frication and (u_anode, v_anode, w_anode) in velocity field components and (P_anode) in pressure.

7.2.2 *Reactions and Flow in Porous Media (Cathode)*

From physics tree click on reacting flow in porous media, and then select transport of concentrated species (rfcs). The species that transport in cathode side are oxygen, nitrogen and water. So, in the table, write (w_O2,w_H2Oc and w_N2) for the mass frication and (u_cathode, v_ cathode, w_ cathode) in velocity field components and (P_ cathode) in pressure; see Fig. 7.3.

Darcy's law is used to study the gas diffusion in gas diffusion layer (GDL) as well as the pore electrodes (Fig. 7.4).

Selection of the type of study: tick the stationary as shown in Fig. 7.5.

Geometrical and Operational Parameter Definitions

Parameter definition:

From the global definition, select parameters, and then fill in the parameters as per shown in Table 7.1.

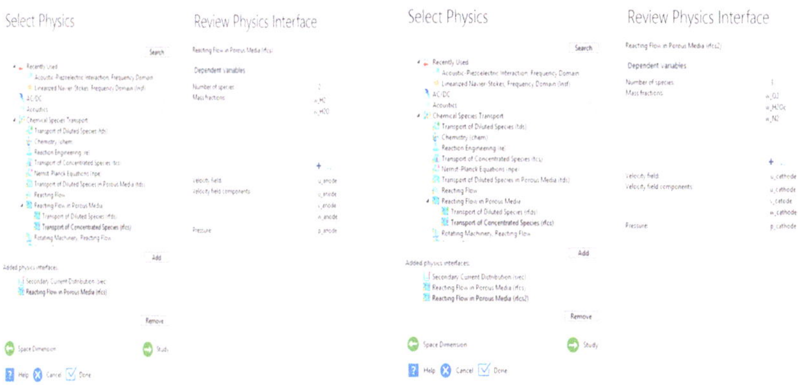

Fig 7.3 Selection of secondary current distribution for both anode and cathode sides

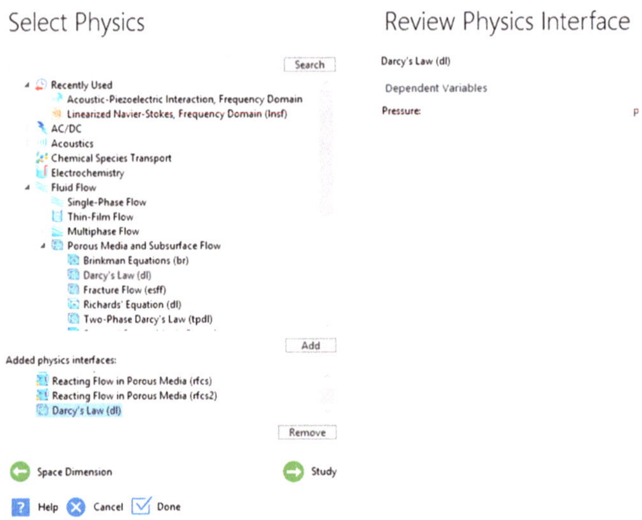

Fig 7.4 Selection of Darcy's law

Fig 7.5 Selection of stationary study

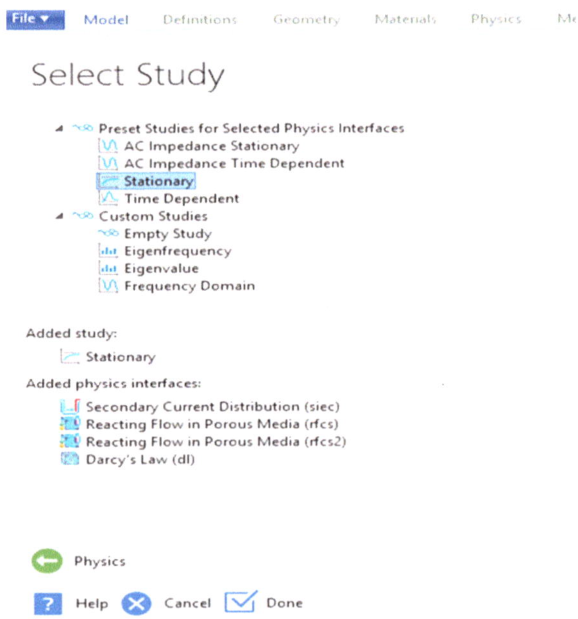

Table 7.1 Geometrical and operational parameters

Parameter's name	Value	Description
C_L	0.06[m]	Cell length
C_W	0.02884528	Cell width
CH_he	1e-3[m]	Channel height
CH_Wid	0.07875[m]	Channel width
RIB_Wid	0.90932e-3[m]	Rib width
GDL_he	0.0002[m]	GDL width
Ele_th	50e-6[m]	Porous electrode thickness
Mem_th	0.00018[m]	Membrane thickness
eps_gdl	0.5	GDL porosity
kappa_gdl	1.79e-11[m^2]	GDL permeability
sigma_gdl	120[S/m]	GDL electric conductivity
wH2_in Inlet H2	0.844	Mass fraction (anode)
wH2O_in Inlet H2O	00.0934	Mass fraction (cathode)
wO2_in Inlet	0.226	Oxygen mass fraction (cathode)
U_in_anode	0.3[m/s]	Anode inlet flow velocity
U_in_cathode	0.3[m/s]	Cathode inlet flow velocity
u_anode	1.19e-5[Pa*s]	Anode viscosity
u_cathode	2.46e-5[Pa*s]	Cathode viscosity
MH2	0.002[kg/mol]	Hydrogen molar mass
MN2	0.028[kg/mol]	Nitrogen molar mass
MH2O	0.018[kg/mol]	Water molar mass
MO2	0.032[kg/mol]	Oxygen molar mass
Deff(H2_H2O)	$6.8815^{-10}*(T)^{1.75}$ [m^2/s]	Binary gas diffusion coefficient (H2_H2O)
Deff(N2_H2O)	$4.0141e\text{-}7*(T])^{1.75}$ [m^2/s]	Binary gas diffusion coefficient
Deff(O2_N2)	$8.5640e\text{-}7*(T)^{1.75}$ [m^2/s]	Binary gas diffusion coefficient
Deff(O2_H2O)	$2.4748e\text{-}7*(T)^{1.75}$ [m^2/s]	Binary gas diffusion coefficient
T	333[K]	Cell temperature
p_ref	1*e5[P]	Reference pressure
V_cell	1.1[v]	Cell voltage
cO2_ref	40.88[mol/m^3]	Oxygen reference concentration
cH2_ref	40.88[mol/m^3]	Hydrogen reference concentration
eps_l 0.3		
eps_cl	1-eps_l-eps_gdl	Open volume fraction for gas diffusion in porous electrodes
kappa_cl	kappa_gdl/5	Permeability (porous electrode)
sigma_m	17.69[S/m]	Membrane conductivity
Re_air	0.55	Relative humidity (air)
Re_Hydrogen	0.55	Relative humidity (hydrogen)

7.3 Building the Geometry Structure of the Model

Geometry of Gas Channel (Anode Side)

The first element that needs to be constructed is the straight gas channel of the anode side. From the geometry section, select the block by right-clicking on geometry, choose the block and then follow the steps shown in Fig 7.6 a, b. (This is the single gas channel; the rest of the gas channel will be created later.)

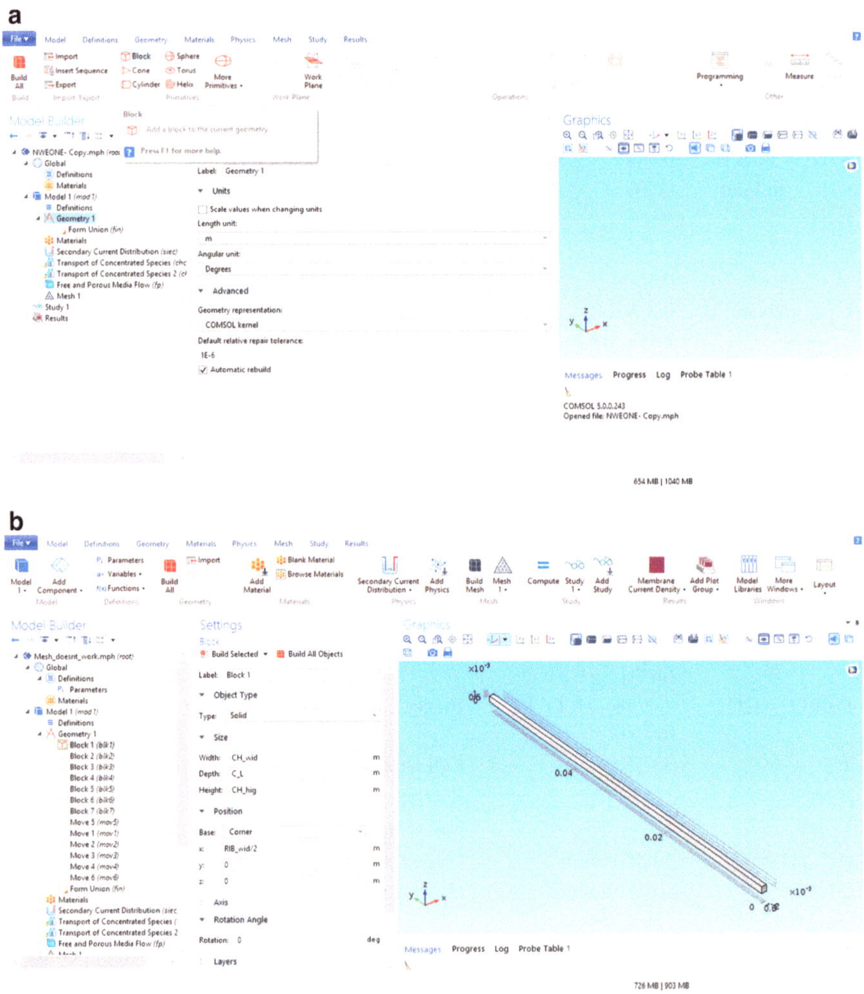

Fig 7.6 (**a**) Selection of block to build up straight gas channel of the anode. (**b**) Shows the setting values of the gas channel

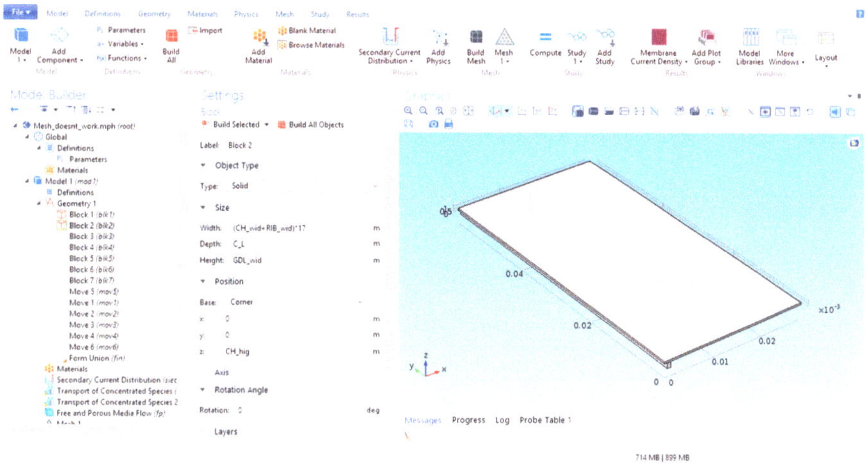

Fig. 7.7 Geometry of GDL

Design of GDL

The second element is the cathode GDL; see Fig. 7.7. Follow the same steps in gas channel to select the block of GDL. Fill the space as follows:

In the width space, type (C_W)
In depth space, type (C_L)
In height space, type (GDL_wid). In position section in the Z space, type (CH_hig).

Geometry of Electrode (Catalyst)

Follow the same steps, for gas channel and GDL, to construct the design shown in Fig. 7.8. In the width space, type (C_W); in the depth space, type(C_L); and in the height space, type (Ele_th). In position section, type (CH_hig+GDL_wid) in Z space.

Geometry Structure of Membrane (Electrolyte)

The membrane is the fourth layer in the cell. Same procedures taken previously will be taken here taking into account the solvency of the blanks in Fig. 7.9.

The Component Geometries of the Cathode Side

Graphical steps for designing electrode at the cathode side, GDL at the cathode side and gas channel at the cathode side are shown in Figs. 7.10, 7.11 and 7.12.

Geometry of the Rest of the Gas Channel for Both the Anode and Cathode Sides

To complete the design and structure of model, the rest of the straight gas channel needs to build up in both sides of a cell.

Geometry of the gas channel, for both sides and complete design, is shown in Fig. 7.13.

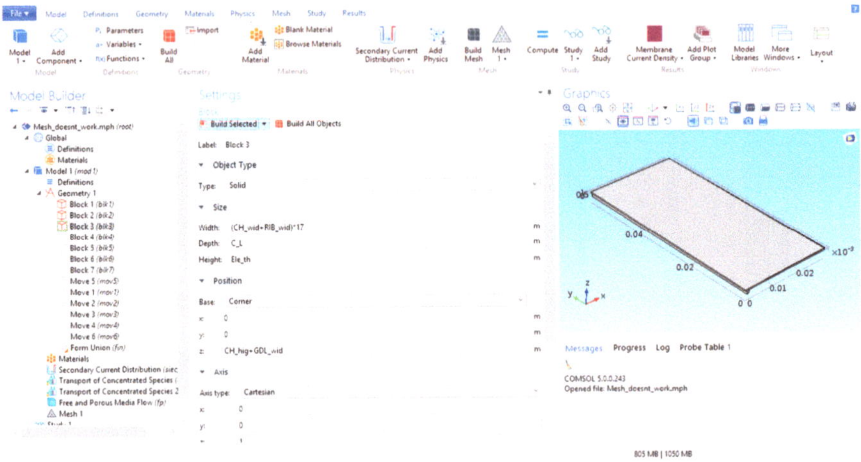

Fig. 7.8 Geometry of electrode

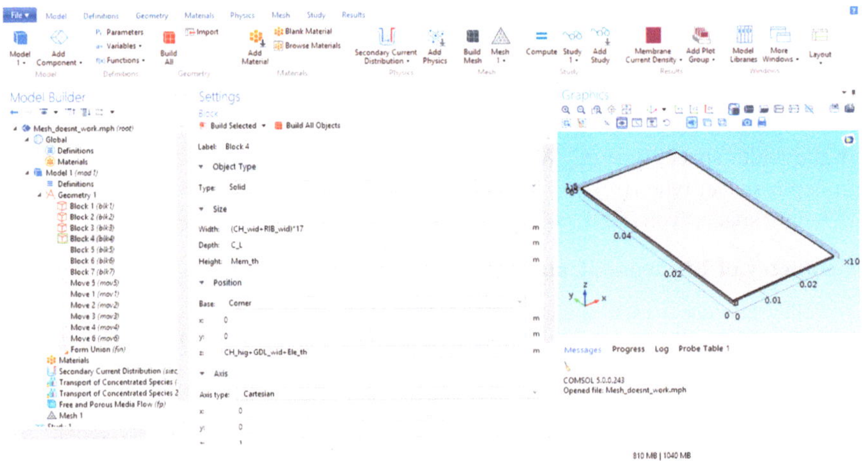

Fig. 7.9 Design of membrane

Meshing Process

The final step in a geometry structure is to finalise by forming a union, uniting all of the geometrical objects and meshing them as a single object. In order to make all layers meshed properly and simultaneously, right-click on Mesh dialog box, select User-controlled mesh and then select Size (see Fig. 7.14).

Fig 7.14: Selection of sequence type of Mesh

Fill the space as shown in Fig. 7.15.

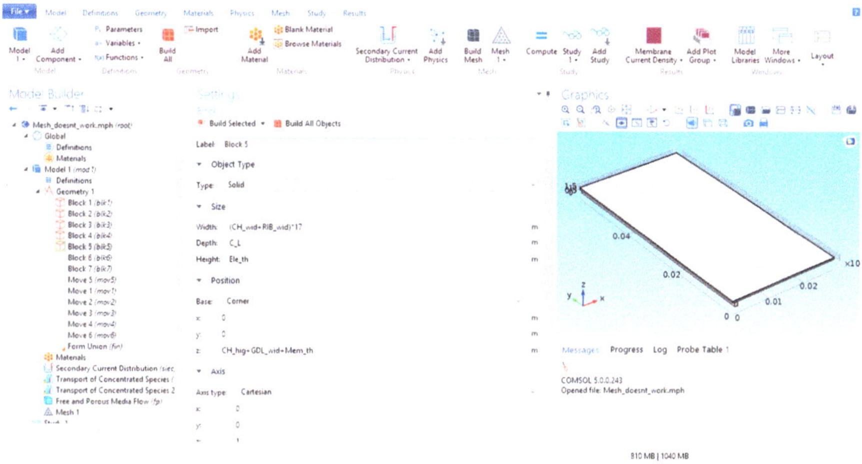

Fig 7.10 Electrode at the cathode side

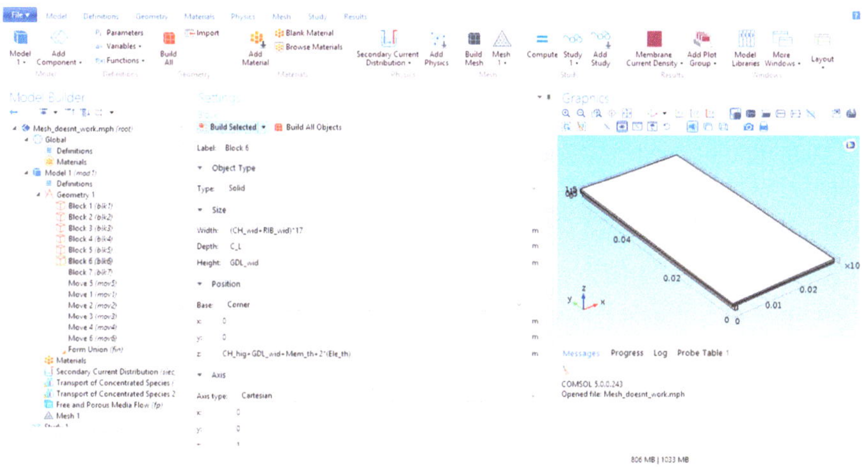

Fig. 7.11 GDL at the cathode side

Definitions

PEM fuel cell consists of three main parts, namely, anode, cathode and membrane. Anode same as a cathode consists of three main layers which are gas channel, gas diffusion layer (GDL) and catalyst layer.

In order to understand the functions of each layer, some definitions need to be specified. This included the input and output of gas channels and the GDL and electrodes.

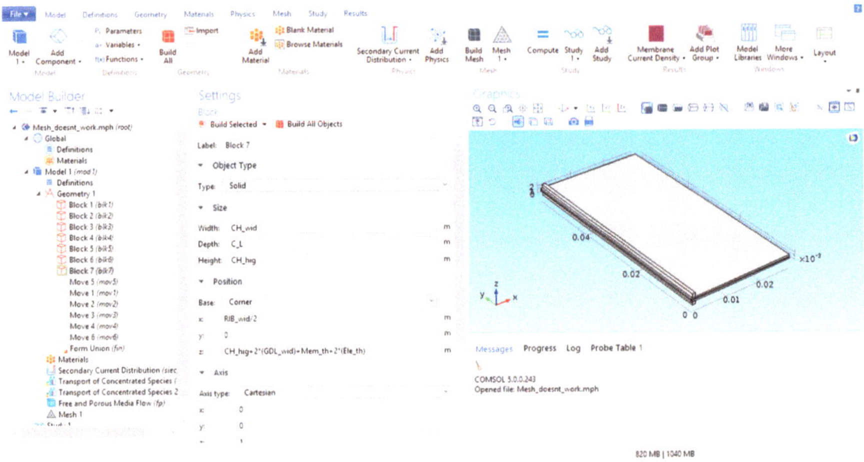

Fig. 7.12 Gas channel at the cathode side

Definition of Anode Gas Channels

1. Right-click on Definition icon, open the Selection list and choose Explicit. To rename the section, right-click and write (Gas_Channel_anode).
2. Same procedures are used to define the GDL and electrode; see Fig. 7.16.

Definition of anode GDL and electrode and inlet and outlet of gas channel are shown in Figs 7.17 and 7.18.

Definition of Membrane

This is shown in Fig. 7.19.

Definition of Cathode Parts

The steps to define cathode components are shown in Fig. 7.20.

Study of the Secondary Current Distribution

Fig 7.21 illustrates how to determine parts of current that passes through anode.

Perouse Electrode (Anode)

Specifications of the electrode are set, using Fig. 7.22 at the following:

1. From the domains in the physics toolbar, select porous electrode.
2. From the porous electrode menu, select porous electrode reaction.
3. In the setting menu, type the parameters shown.

Final model that shows transports of concentrated species in anode and cathode sides is shown in Figs 7.23, 7.24 and 7.25.

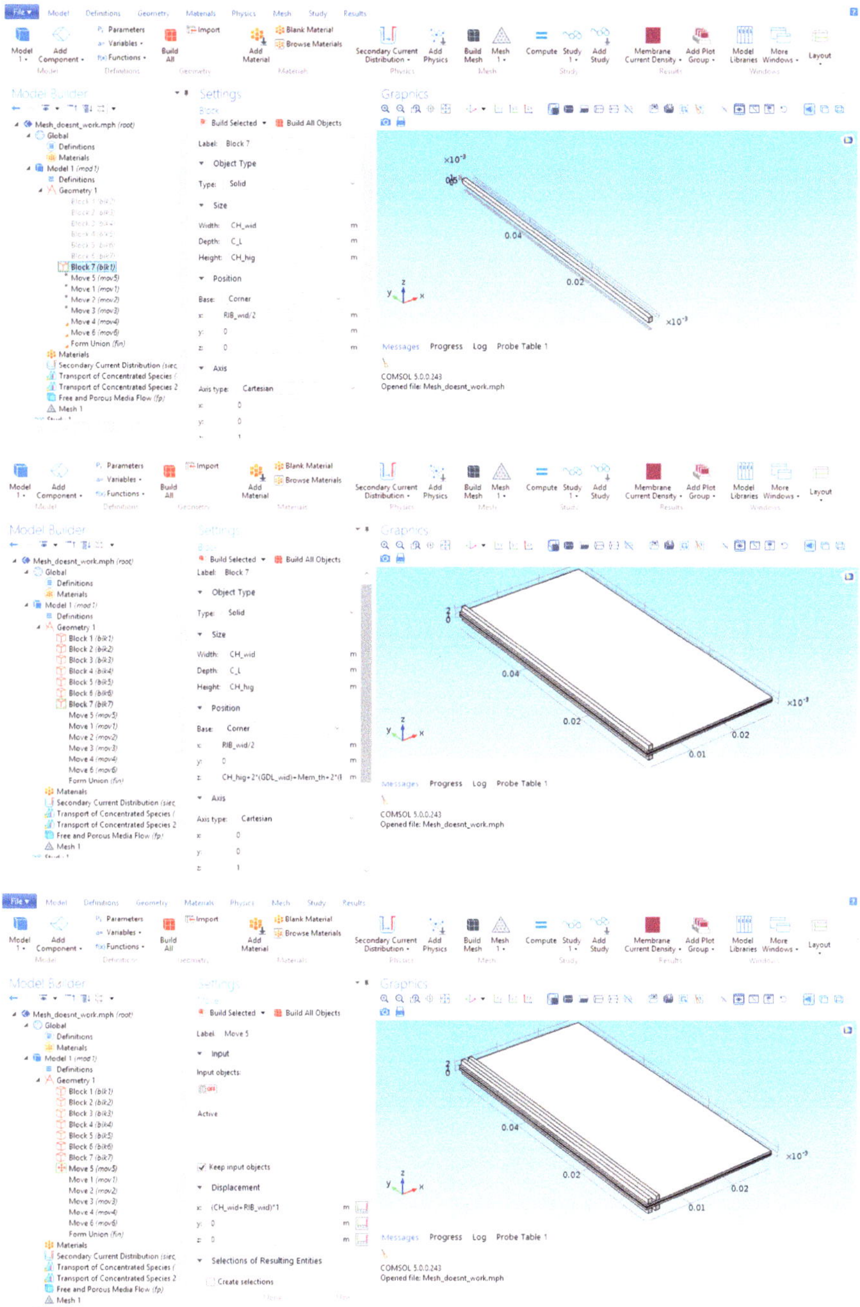

Fig. 7.13 Geometry of the gas channel for both sides and complete design

Fig. 7.13 (continued)

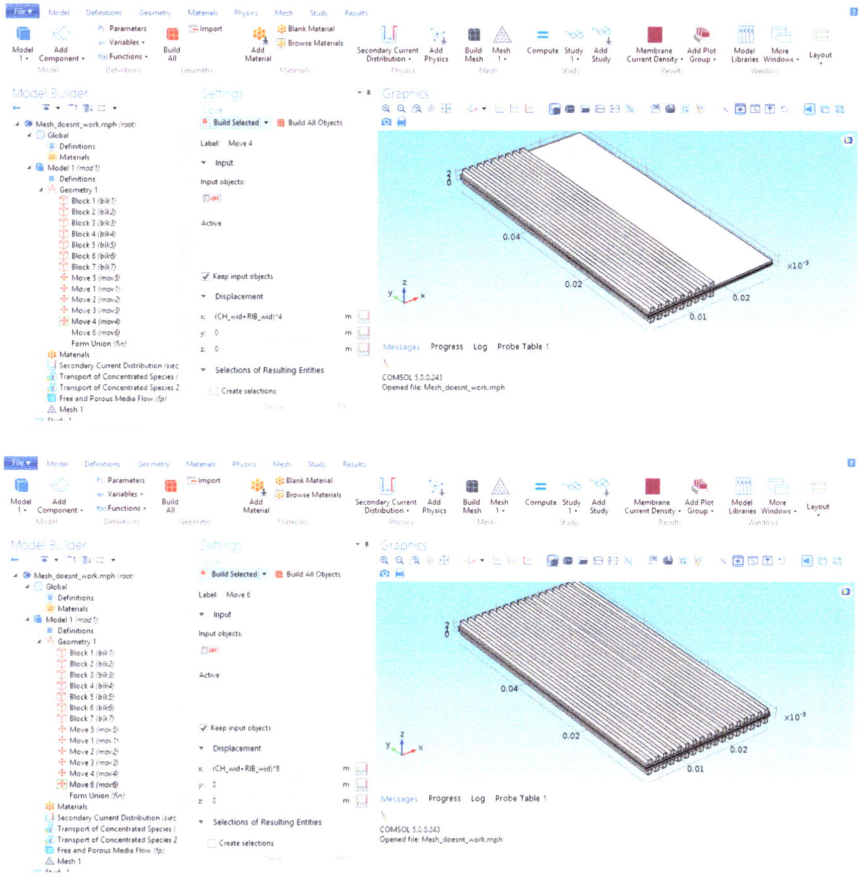

Fig. 7.13 (continued)

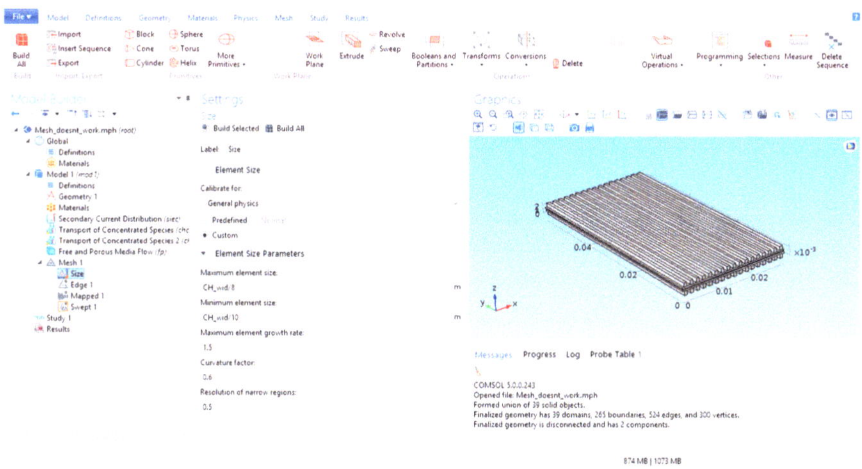

Fig 7.14 Selection of size setting

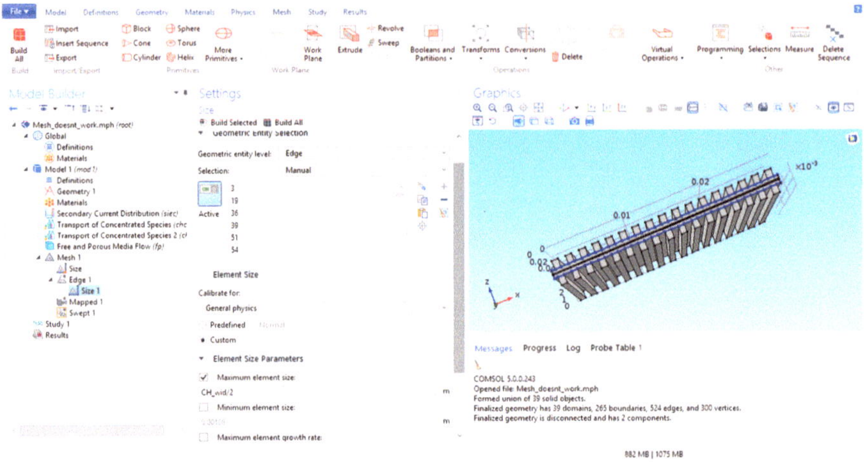

Fig. 7.15 Mesh of the GDL edge

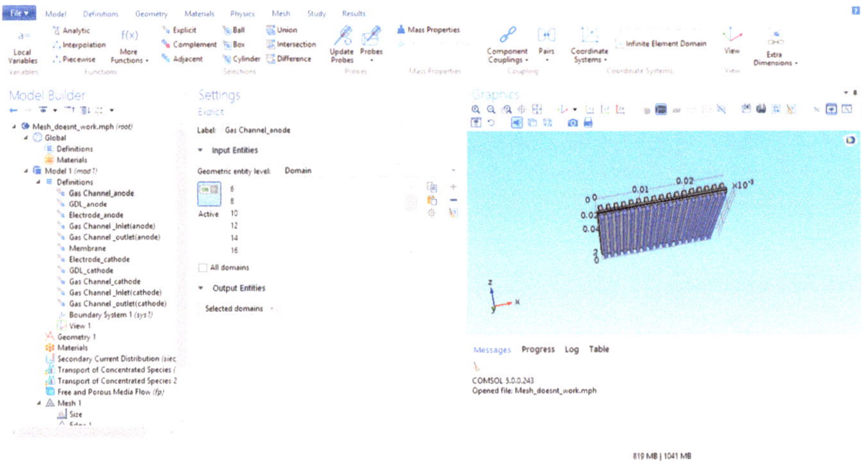

Fig. 7.16 Definition of inlet and outlet of gas channel

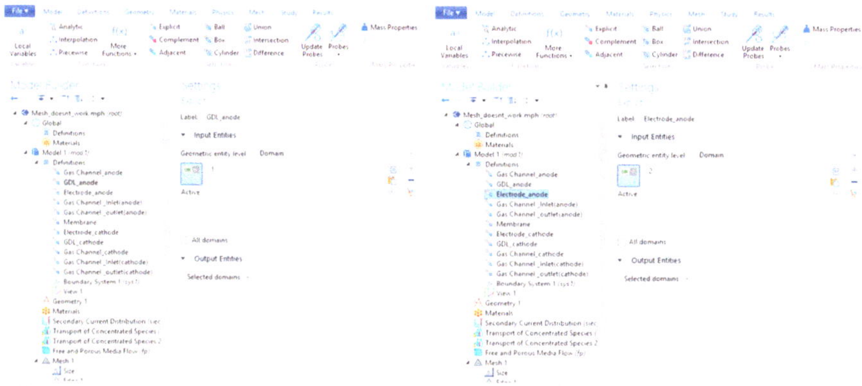

Fig. 7.17 Definition of anode GDL and electrode

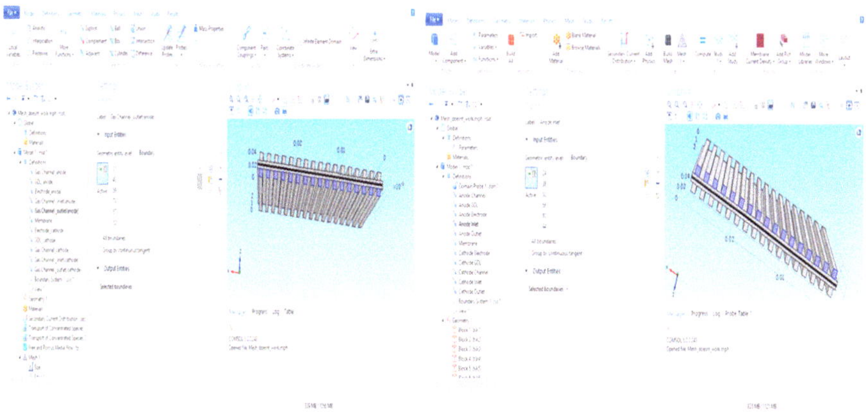

Fig. 7.18 Definition of anode inlet and outlet of gas channel

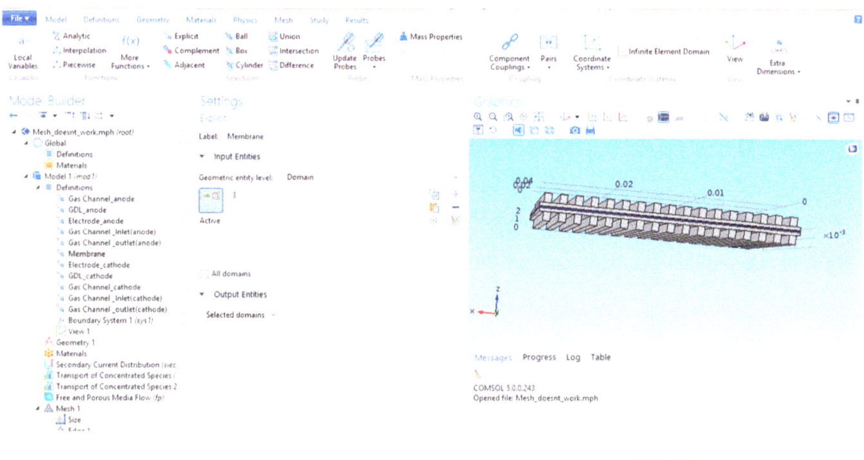

Fig. 7.19 Definition of membrane

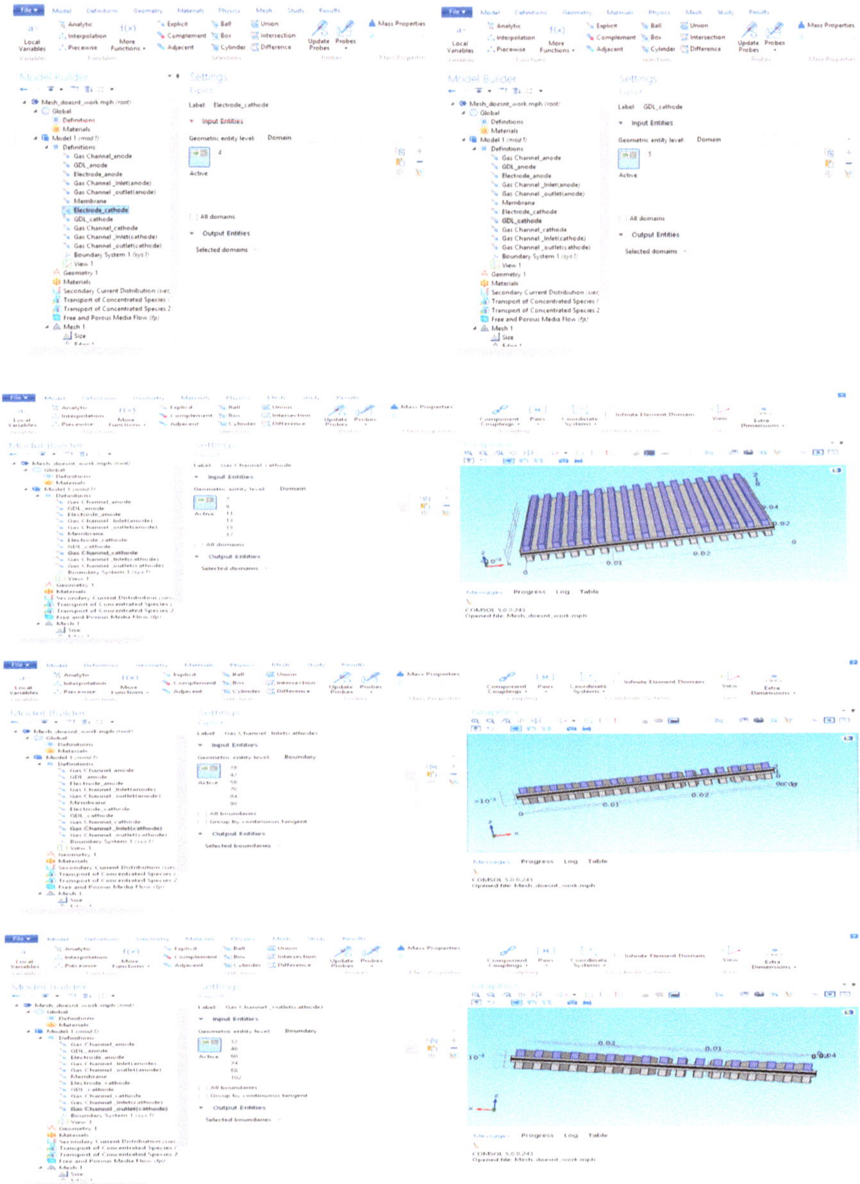

Fig 7.20 Definition of cathode components

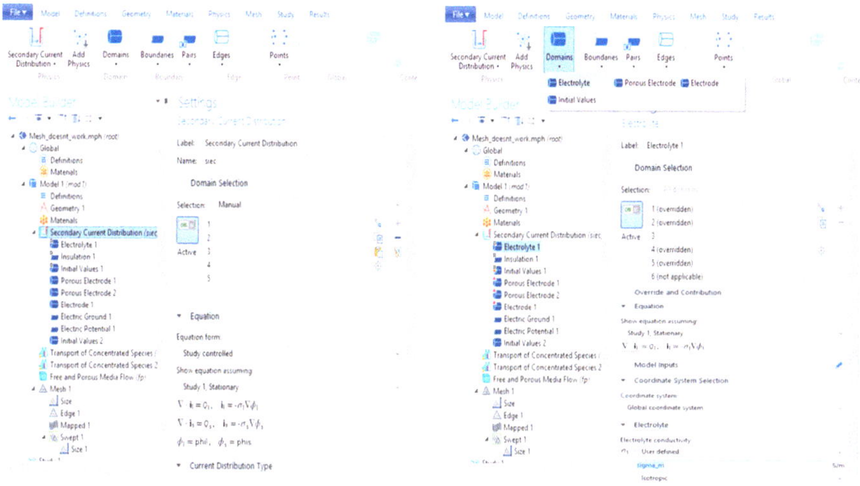

Fig. 7.21 Setting of secondary current distribution

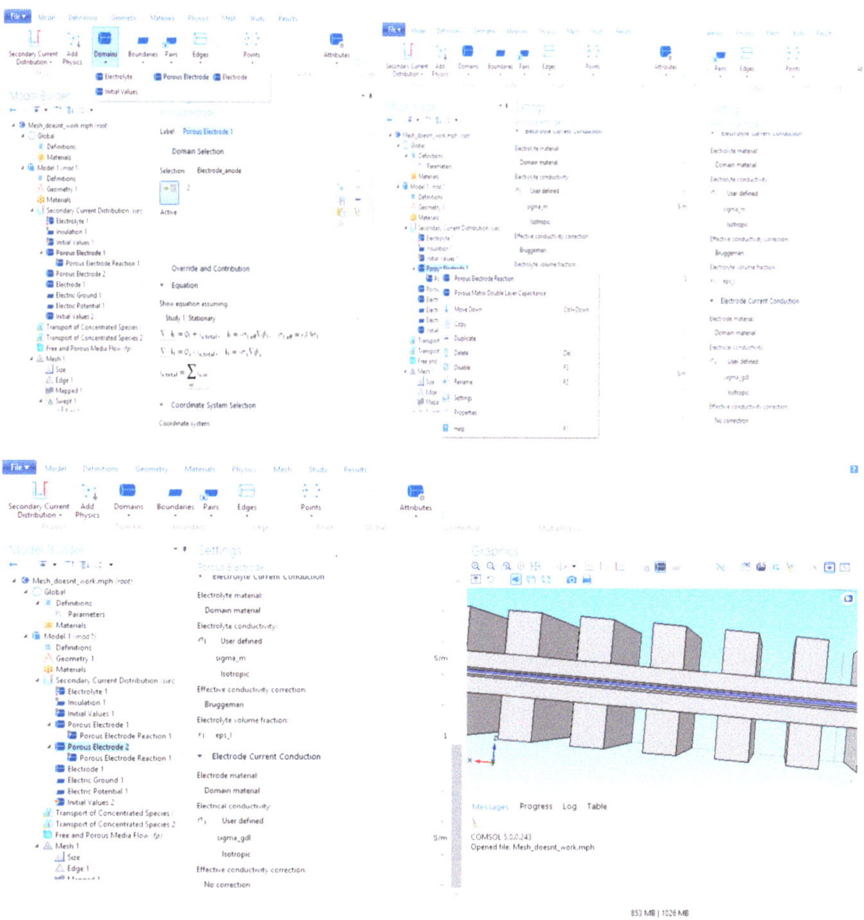

Fig. 7.22 The parameter and specification setting of electrode

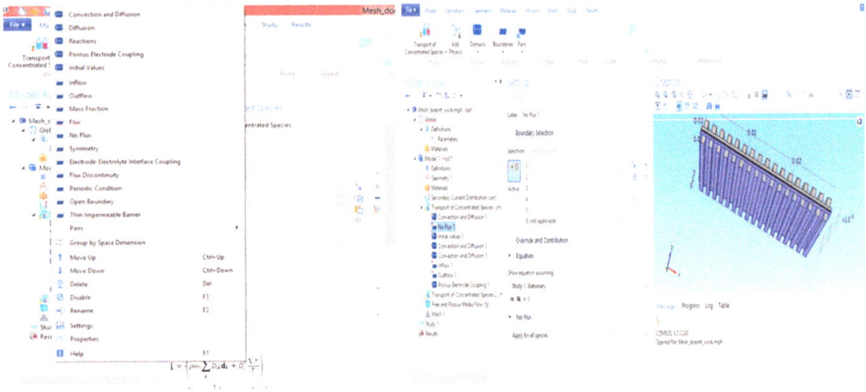

Fig 7.23 Transport of concentrated species in anode side

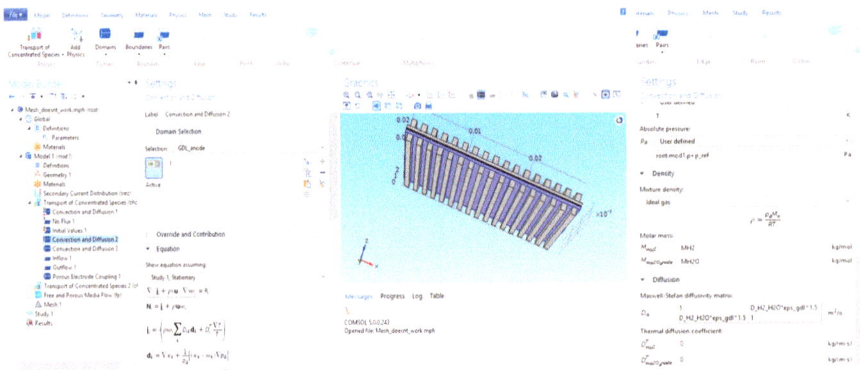

Fig 7.24 Transport of concentrated species in anode side

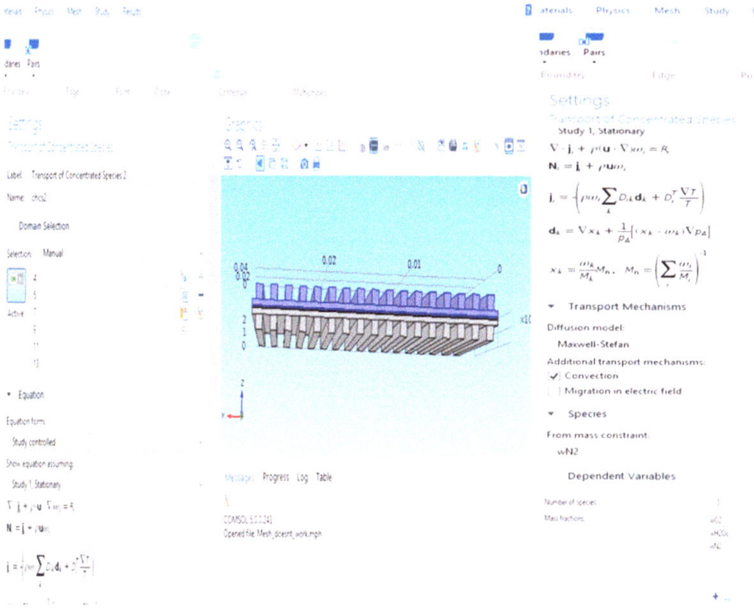

Fig 7.25 Transport of concentrated species in cathode side

Right-click on Size and select Edge. In the setting section, select the upper edge of GDL for both sides as shown in Figs. 7.26 and 7.27.

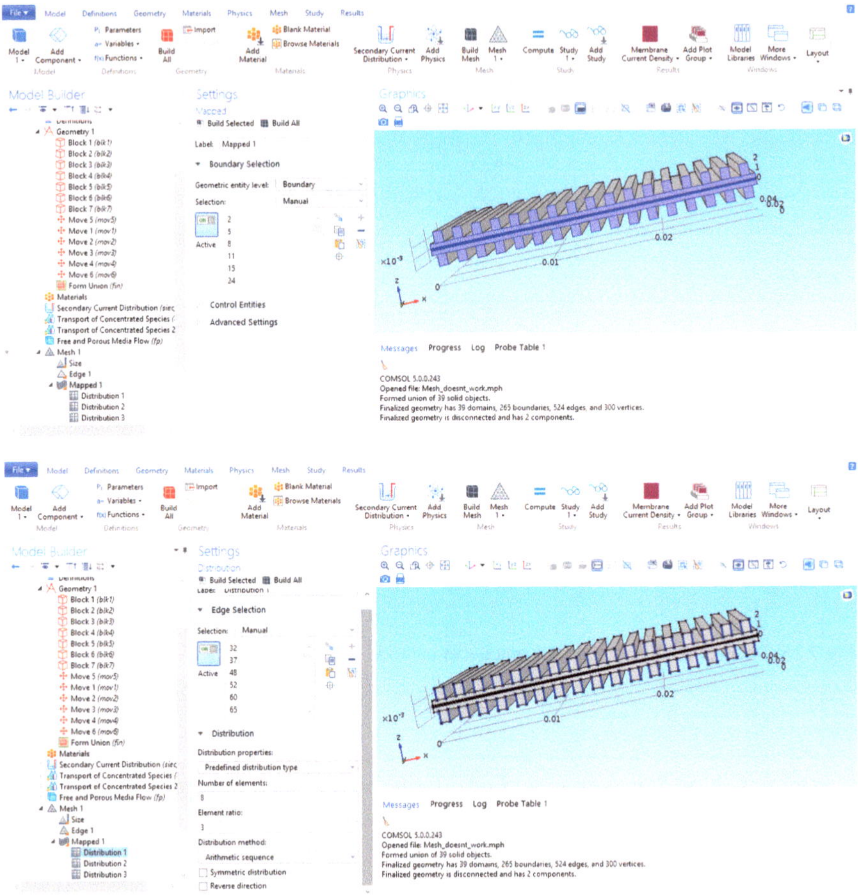

Fig. 7.26 Map of the boundaries

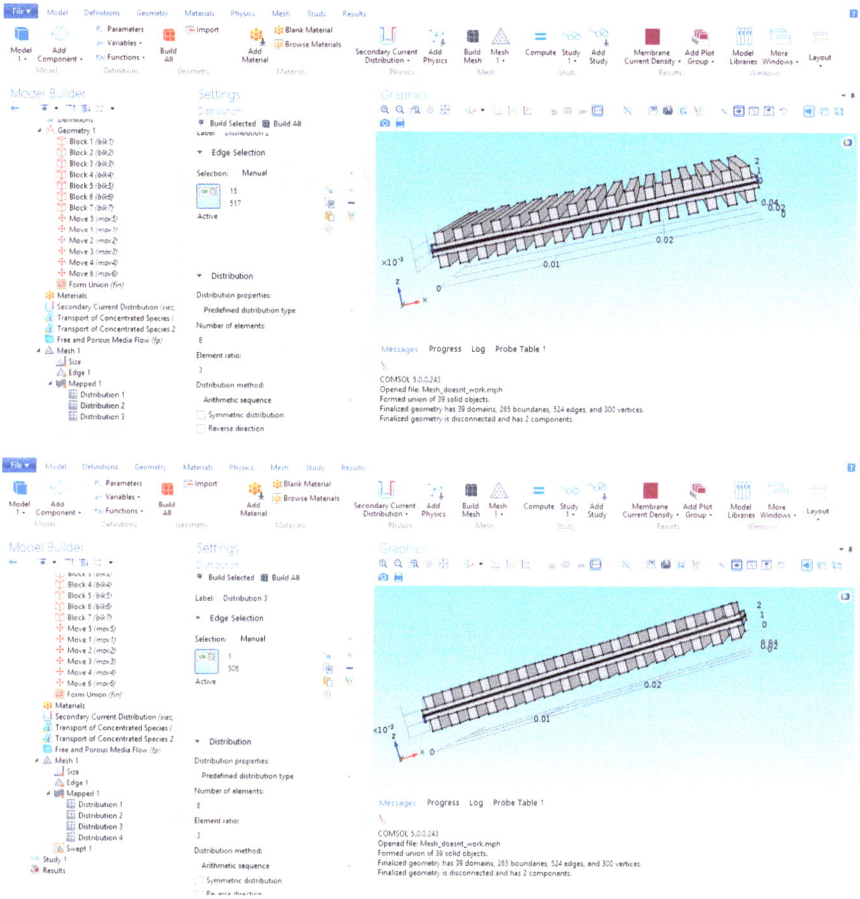

Fig. 7.26 (continued)

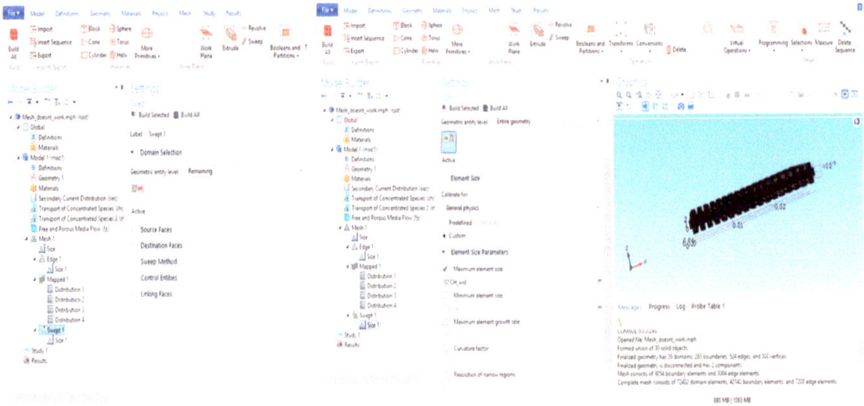

Fig 7.27 Meshed model

References

1. Alrweq, M., Albarbar, A. (2016). *Investigation into the characteristics of proton exchange membrane fuel cell-based power system IET science, measurement & technology*. doi: https://doi.org/10.1049/iet-smt.2015.0046 , Online ISSN 1751–8830.
2. Albarbar, A., Alrweq, M. (2016). *Proton exchange membrane fuel cells: recent developments, modelling and challenges*. Proceeding of 6th International Symposium in Energy Challenges and Mechanics, Inverness.
3. Albarbar, A., Alrweq, M. (2017). *Acoustic emission based condition monitoring technique for fuel cell systems*. Proceeding of 7th International Symposium in Energy Challenges and Mechanics, Manchester.

Appendix A: Parameters that Used to Model PEM Fuel Cells

Name	Value	Description
L	0.06[m]	Cell length
H_ch	1e-3[m]	Channel height
W_ch	9.474e-3[m]	Channel width
W_rib	9.0932e-3[m]	Rib width
H_gdl	640e-6[m]	GDL width
H_electrode	50e-6[m]	Porous electrode thickness
H_membrane	100e-6[m]	Membrane thickness
eps_gdl	0.4	GDL porosity
kappa_gdl	1.18e-11[m^2]	GDL permeability
sigma_gdl	222[S/m]	GDL electric conductivity
wH2_in	0.743	Inlet H2 mass fraction (anode)
wH2O_in	0.023	Inlet H2O mass fraction (cathode)
wO2_in	0.228	Inlet oxygen mass fraction (cathode)
U_in_anode	0.2[m/s]	Anode inlet flow velocity
U_in_cathode	0.5[m/s]	
mu_anode	1.19e-5[Pa*s]	Anode viscosity
mu_cathode	2.46e-5[Pa*s]	Cathode viscosity
MH2	0.002[kg/mol]	Hydrogen molar mass
MN2	0.028[kg/mol]	Nitrogen molar mass
MH2O	0.018[kg/mol]	Water molar mass
MO2	0.032[kg/mol]	Oxygen molar mass
D_H2_H2O	9.15e-5*(T/307.1[K])$^{1.75}$ [m^2/s]	H2-H2O
D_N2_H2O	2.56e-5*[T/307.15[K])$^{1.75}$ [m^2/s]	N2-H2
D_O2_N2	2.2e-5*[T/293.2[K])$^{1.75}$ [m^2/s]	O2-N2
D_02_H2O	2.82e-5*(T/308.1[K])$^{1.75}$ [m^2/s]	O2-H2O

(continued)

A. Albarbar, M. Alrweq, *Proton Exchange Membrane Fuel Cells*,
https://doi.org/10.1007/978-3-319-70727-3

Name	Value	Description
T	20+273.15[K]	Cell temperature
p_ref	101e3[Pa]	Reference pressure
V_cell	0.9	Cell voltage
cO2_ref	40.88[mol/m^3]	Oxygen reference concentration
cH2_ref	40.88[mol/m^3]	Hydrogen reference concentration
eps_l	0.3	
eps_cl	1-eps_l-eps_gdl	Open volume fraction for fusion in porous electrodes
kappa_cl	kappa_gdl/5	Permeability (porous electrodes)
sigma_m	9.825[s/m]	Membrane conductivity

Appendix B: Current Distribution and Water Concentration in GDL

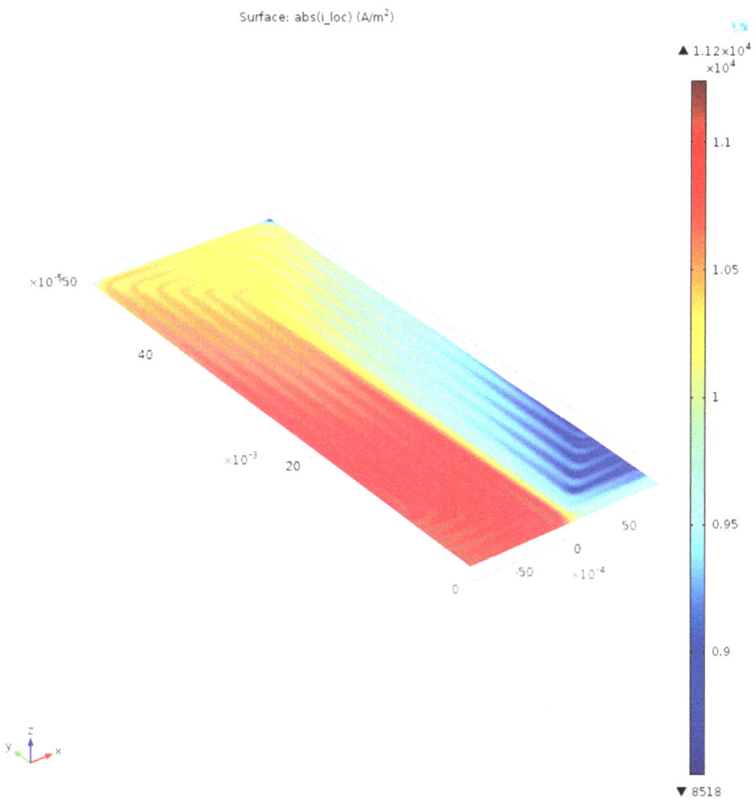

Current density in GDL at operating conditions (298.15 K and 1.5 atm)

A. Albarbar, M. Alrweq, *Proton Exchange Membrane Fuel Cells*,
https://doi.org/10.1007/978-3-319-70727-3

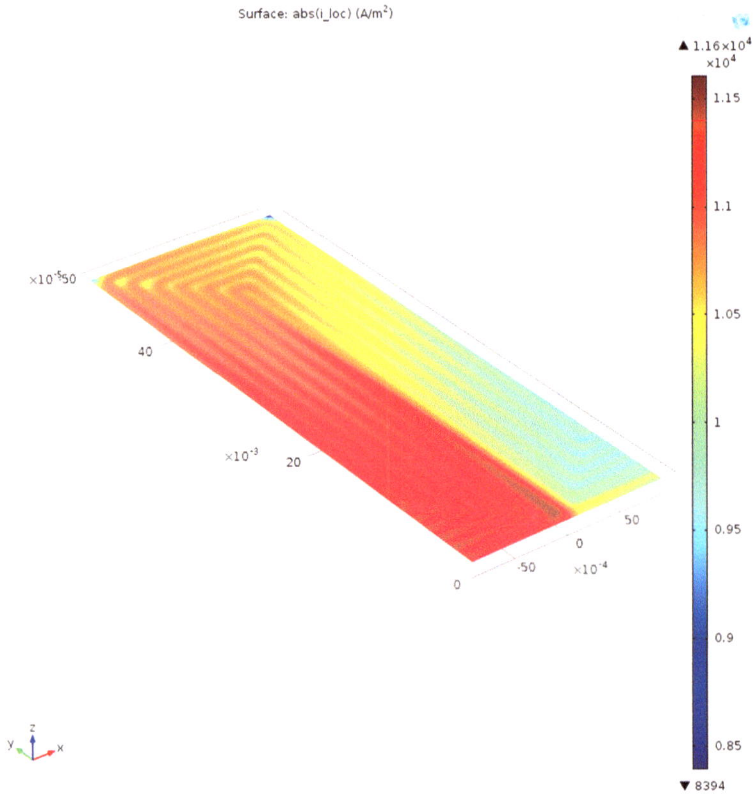

Current density in GDL at operating conditions (323.15 K and 1.5 atm)

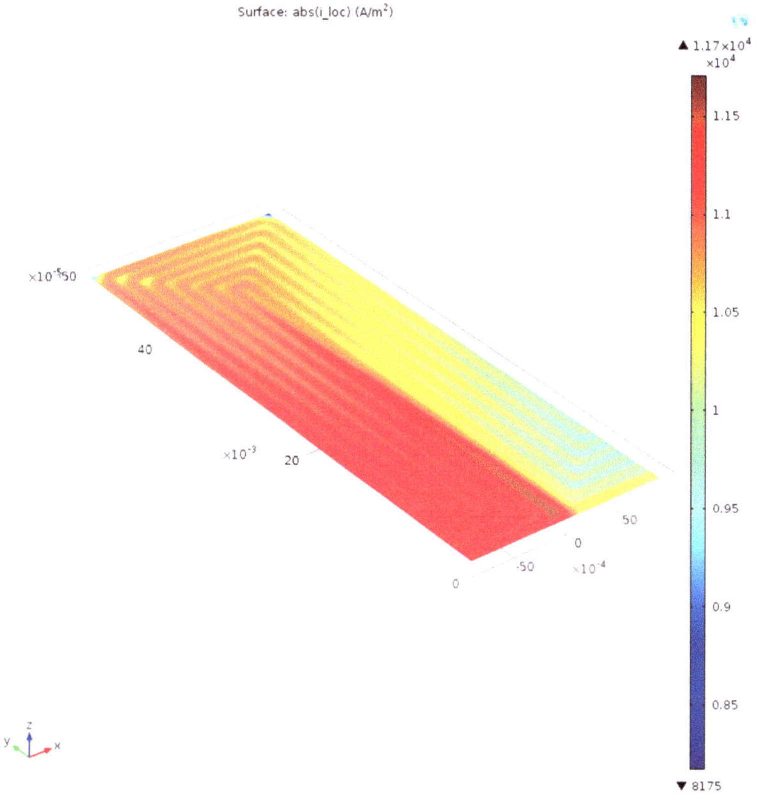

Current density in GDL at operating conditions (348.15 K and 1.5 atm)

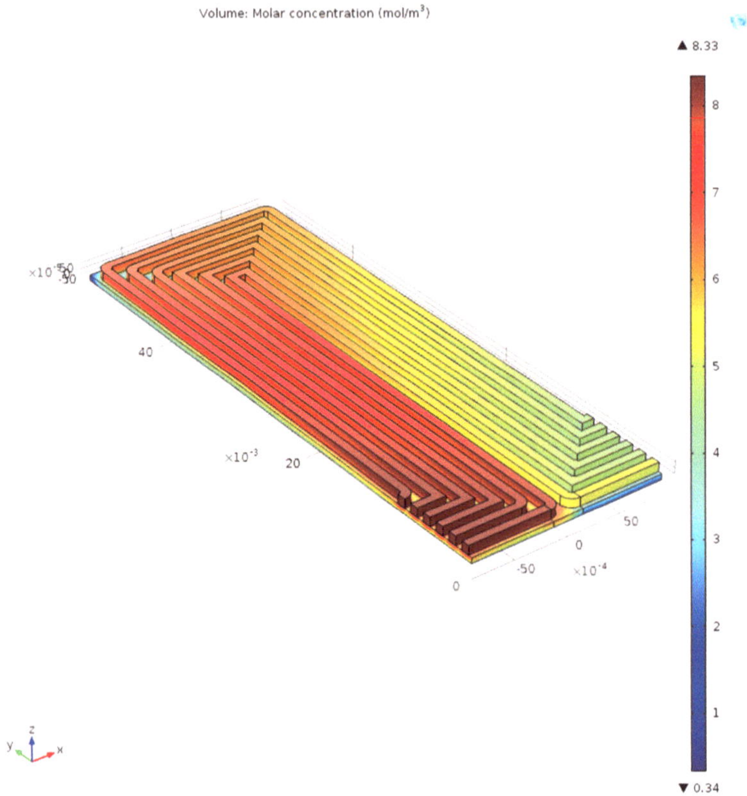

Water concentration in GDL (298.15 K and 1.5 atm)

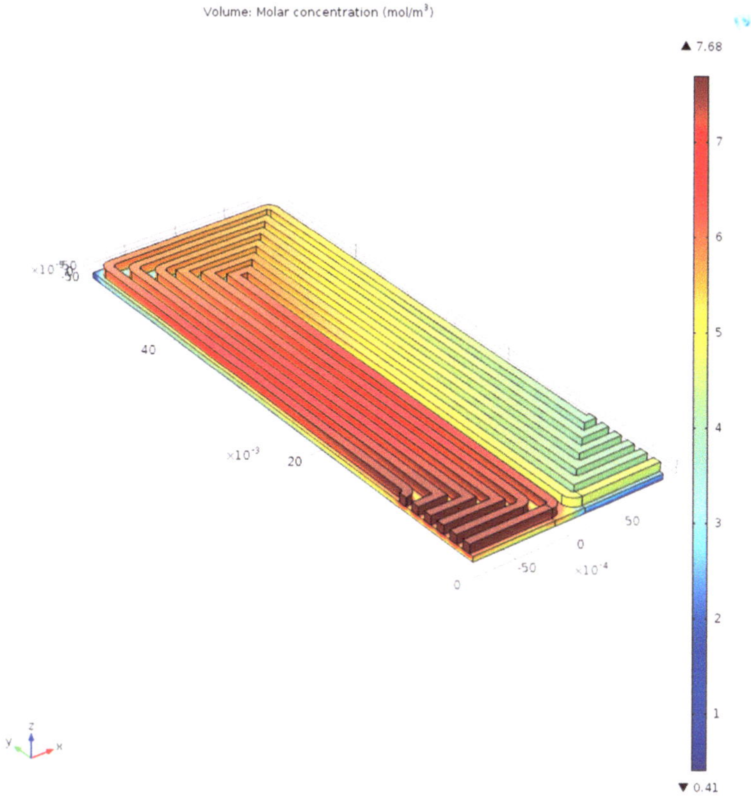

Water concentration in GDL (323.15 K and 1.5 atm)

Appendix C: Current Density in Membrane

A. Albarbar, M. Alrweq, *Proton Exchange Membrane Fuel Cells*,
https://doi.org/10.1007/978-3-319-70727-3

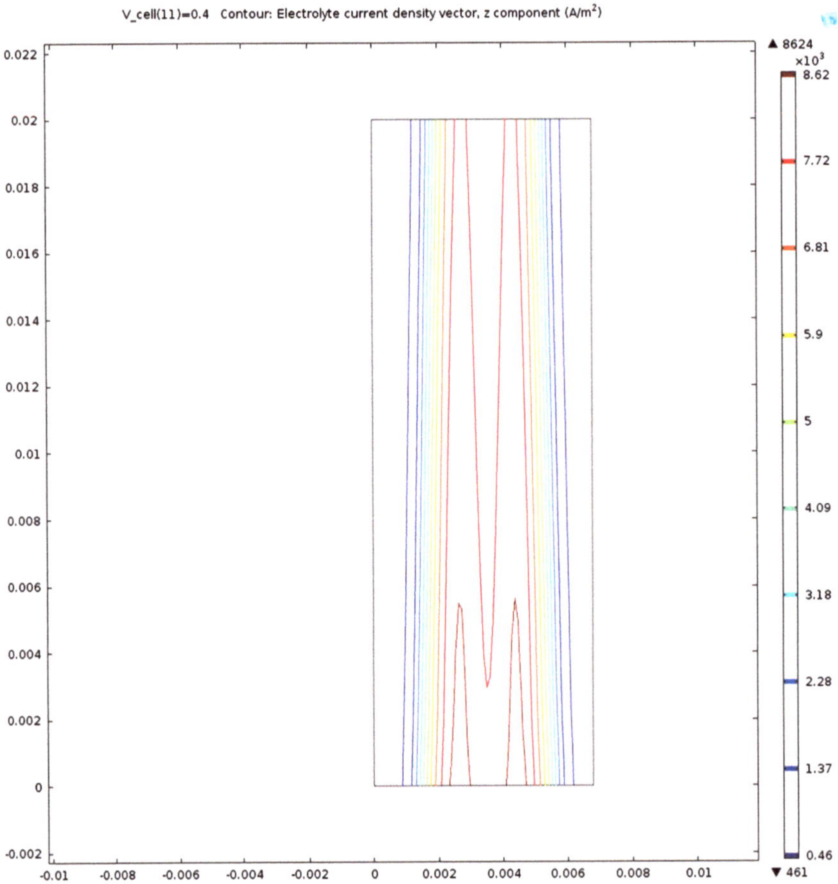

Current density distribution in membrane (298.15 K and 1.5 atm)

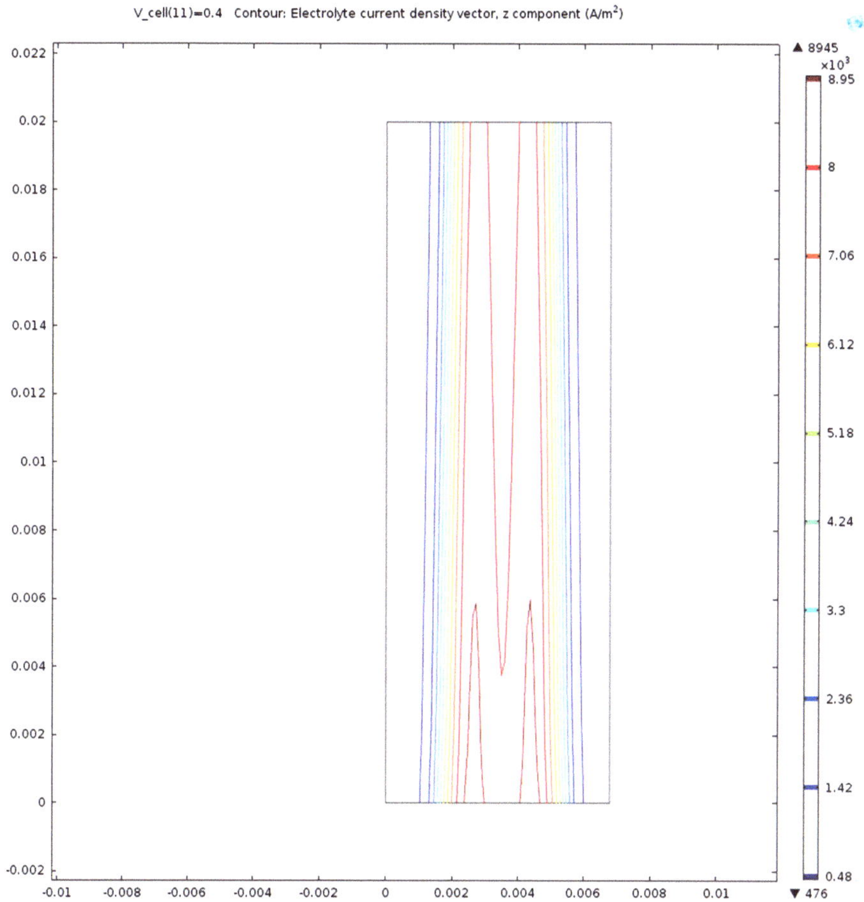

Current density distribution in membrane (223.15 K and 1.5 atm)

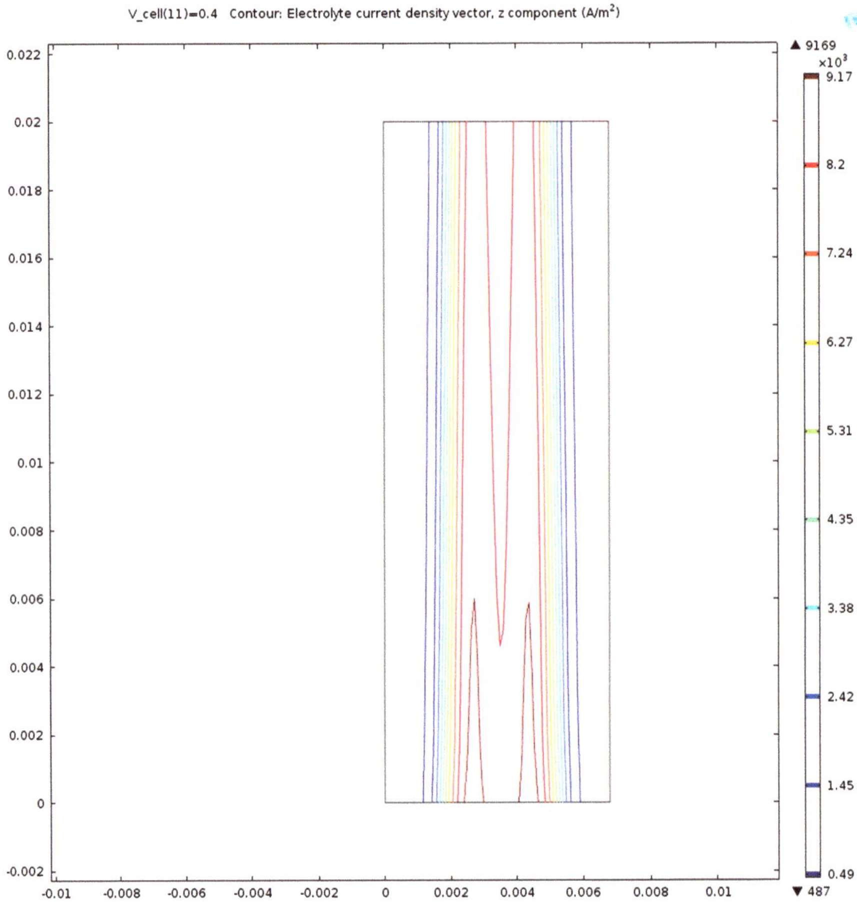

Current density distribution in membrane (248.15 K and 1.5 atm)

Index

A. Albarbar, M. Alrweq, *Proton Exchange Membrane Fuel Cells*,
https://doi.org/10.1007/978-3-319-70727-3

G

H

K

M

N

O

MIX
Papier aus verantwortungsvollen Quellen
Paper from responsible sources
FSC® C105338

If you have any concerns about our products,
you can contact us on
ProductSafety@springernature.com

In case Publisher is established outside the EU,
the EU authorized representative is:
Springer Nature Customer Service Center GmbH
Europaplatz 3, 69115 Heidelberg, Germany

Printed by Libri Plureos GmbH
in Hamburg, Germany